Spinoza's Philosophy of Ratio

Edited by Beth Lord

EDINBURGH
University Press

Edinburgh University Press is one of the leading university presses in the UK. We publish academic books and journals in our selected subject areas across the humanities and social sciences, combining cutting-edge scholarship with high editorial and production values to produce academic works of lasting importance. For more information visit our website: edinburghuniversitypress.com

© editorial matter and organisation Beth Lord, 2018
© the chapters their several authors, 2018

Edinburgh University Press Ltd
The Tun – Holyrood Road
12(2f) Jackson's Entry
Edinburgh EH8 8PJ

Typeset in 10/12 Goudy Old Style by
Servis Filmsetting Ltd, Stockport, Cheshire

A CIP record for this book is available from the British Library

ISBN 978 1 4744 2043 3 (hardback)
ISBN 978 1 4744 2044 0 (webready PDF)
ISBN 978 1 4744 2045 7 (epub)

The right of Beth Lord to be identified as the editor of this work has been asserted in accordance with the Copyright, Designs and Patents Act 1988, and the Copyright and Related Rights Regulations 2003 (SI No. 2498).

Contents

Acknowledgements v
Abbreviations of Spinoza's Works vi

Introduction 1
Beth Lord

1 Spinoza's Ontology Geometrically Illustrated: A Reading of
   Ethics IIP8S 5
   Valtteri Viljanen

2 Reason and Body in Spinoza's Metaphysics 19
   Michael LeBuffe

3 Ratio and Activity: Spinoza's Biologising of the Mind in an
   Aristotelian Key 33
   Heidi M. Ravven

4 Harmony in Spinoza and his Critics 46
   Timothy Yenter

5 Ratio as the Basis of Spinoza's Concept of Equality 61
   Beth Lord

6 Proportion as a Barometer of the Affective Life in Spinoza 74
   Simon B. Duffy

7 Spinoza, Heterarchical Ontology, and Affective Architecture 89
   Gökhan Kodalak

8 Dissimilarity: Spinoza's Ethical Ratios and Housing Welfare 108
   Peg Rawes

9 The Greater Part: How Intuition Forms Better Worlds 125
   Stefan White

10 Slownesses and Speeds, Latitudes and Longitudes: In the Vicinity of
   Beatitude 141
   Hélène Frichot

11 The Eyes of the Mind: Proportion in Spinoza, Swift, and Ibn Tufayl   155
    Anthony Uhlmann

Notes on Contributors   169
Bibliography   172
Index   185

Acknowledgements

This book is an outcome of *Equalities of Wellbeing in Philosophy and Architecture*, a research project funded by the Arts and Humanities Research Council (www.ahrc.ac.uk) in 2013 to 2016. Led by Beth Lord and Peg Rawes, *Equalities of Wellbeing* (www.equalitiesofwellbeing.co.uk) aimed to investigate the relevance of Spinoza's thought to architecture and housing design through a series of workshops, publications, and a film. The film, *Equal by Design*, uses Spinoza to focus on inequality, wellbeing, and the lack of availability of well-designed affordable housing in the UK. It is very much a companion to this book and is free to view at www.equalbydesign.co.uk. The chapters in this book were presented at the project's 'Spinoza and Proportion' conference at the University of Aberdeen. The editor and authors are grateful to the AHRC for its support. The editor would like to thank Peg Rawes, the project's co-investigator, and Christopher Thomas who, as a PhD student, worked to support the project and prepared the bibliography for this book.

Wall Drawing #122 by Sol LeWitt (detail) is reproduced with the permission of Paula Cooper Gallery, the Design and Artists Copyright Society, and the LeWitt estate.

Abbreviations of Spinoza's Works

CGH   *Hebrew Grammar*; references are to chapter number and to volume and page number in G

CM   *Metaphysical Thoughts* (appendix to PPC); references are to part number in roman numerals and chapter number in arabic numerals

E   *Ethics*; references are to part number in roman numerals, followed by Proposition (Definition, Axiom etc.) number in arabic numerals, as follows: D = Definition; A = Axiom; P = Proposition; Dem. = Demonstration; C = Corollary; S = Scholium; Exp. = Explanation; L = Lemma; Post. = Postulate; Pref. = Preface; App. = Appendix; Def.Aff. = Part III 'Definitions of the Affects' (e.g. E IVP37S2 = *Ethics* Part IV, Proposition 37, Scholium 2)

Ep.   Letters; references are to letter number, correspondent and date (where known)

G   *Spinoza Opera*, ed. Carl Gebhardt, 4 vols (1925)

KV   *Short Treatise on God, Man, and his Wellbeing*; references are to part number in roman numerals and chapter number in arabic numerals

PPC   *Principles of Cartesian Philosophy*; references follow the same system as the *Ethics*

TIE   *Treatise on the Emendation of the Intellect*; references are to paragraph number

TP   *Political Treatise*; references are to chapter and paragraph number

TTP   *Theological-Political Treatise*; references are to chapter number, and to volume and page number in G

Introduction

Beth Lord

This book is about Spinoza's philosophy of ratio. The Latin term *ratio* can mean reason, relation, and proportion, as well as mathematical ratio. It is all these senses of *ratio*, and the relations between them, that we address in this book. The book argues that Spinoza's philosophy is a *philosophy of ratio*: not a 'rationalist' philosophy, but a philosophy based on the interactions of reason, relation, and proportion. In this short introduction I introduce these concepts, direct the reader to the chapters that discuss them, and consider how Spinoza's philosophy of *ratio* is reflected in architecture, one of the book's key themes.

Ratio is a significant term in Spinoza's philosophy. Spinoza is typically (if not entirely accurately) characterised as a rationalist, for whom reasoning has central importance. In the *Ethics*, reason is Spinoza's 'second kind' of knowledge (E IIP40S2), sitting between empirical awareness or imagination (knowledge of the first kind) and intuitive intellection (knowledge of the third kind). In reasoning, we understand things adequately: we start from axioms, definitions, and basic properties and analytically or deductively build up true understanding of the causes of – or reasons for – things being as they are. Reasons are what reason understands; as Michael LeBuffe discusses in Chapter 2, those reasons may be ideal or corporeal. Developing one's reasoning is the primary goal of the human mind, and is key to our flourishing: the ethical arguments of Spinoza's *Ethics* rest on the principle that our wellbeing, virtue, and freedom develop in tandem with our reasoning. The freest and ethically 'best' person – that is, the person who has the most autonomous control of her or his own actions and reactions – is also the most rational.

But reason is only the most obvious sense of *ratio* in Spinoza's philosophy. Spinoza says that every physical body is governed by a characteristic 'ratio of motion and rest' (E IIL5). This may be understood as a mathematical ratio (of degrees of motion to degrees of rest) but Heidi M. Ravven, in Chapter 3, argues that it may be better understood as the body's unique equilibrium. This characteristic ratio determines and provides the reason for the body's individual form: as LeBuffe argues, bodies contain their own reasons. In striving to understand these ratios/reasons, the mind strives to maintain its own equilibrium which, according to Ravven, Spinoza conceives as 'biologised'. *Ratio* underlies Spinoza's parallelism doctrine, and focusing on it allows new interpretations of that doctrine to be developed.

Ratio determines what a body is, and what the mind can understand about that body. Yet for Spinoza, the body is not determined exclusively by its characteristic ratio. The human body, for example, can maintain its form, grow, and flourish only through its interactions with food, water, shelter, tools, and other human, animal, and inanimate bodies. These interactions are relations – *ratio* in its third sense – and are essential to finite existence. Finite things are defined by being determined by other finite things (E ID5, IP28): our existence, both physical and mental, necessarily involves relations. The body is both a whole of interrelating parts, and a relational part of larger wholes. In Chapter 5 I argue that Spinoza understands parts of wholes to have ratios that are geometrically equal. Members of communities and states can also be understood to have geometrically equal ratios, giving way to a Spinozan ideal of the equal society based not on political equality, but on equality of flourishing.

While relationality is crucial to our being and flourishing, however, it also threatens us: our interactions can be physically and mentally harmful, and can affect us with negative emotions, leading to a diminishment of reasoning and wellbeing. Relations, good and bad, determine our emotional and social lives. In Chapter 6, Simon B. Duffy examines two ways of interpreting the power to act of Spinoza's relational individual. While both Gilles Deleuze and Pierre Macherey understand that power in terms of ratio, whether that ratio is variable or fixed determines our understanding of how we are affected by other things. In Chapter 7, Gökhan Kodalak considers the agency of built and natural environments in affecting our emotional and social lives. Hélène Frichot focuses on joyful affects in Chapter 10, considering how specific relations in artistic practice can push us towards beatitude, the highest state of freedom, which Spinoza discusses in *Ethics* Part V.

Ratio, then, is a key concept in Spinoza's theories of knowledge, bodies, the affects, and politics. *Ratio* also plays a role in Spinoza's metaphysics, through the concept of proportion. Despite his denial of divine will and final purposes, Spinoza appears committed to the belief that the universe is proportional and harmonious. This seeming paradox is discussed by Timothy Yenter in Chapter 4, and is the starting point for considering the many senses of 'harmony' available to Spinoza. In Chapter 11, Anthony Uhlmann uses the Islamic philosopher Ibn Tufayl and the English novelist Jonathan Swift as critical lenses through which to examine proportionality in Spinoza's work. The proportionality of the universe underlies the notion that there are 'good' proportions – in human bodies, minds, relationships, and communities – that we aspire to realise as we become more rational; an idea satirised by Swift. For Spinoza, these proportions are embedded in the essences of things, and can be well or badly understood, and well or badly realised.

The highly rational person is able to understand the essences of things and the ratios, relations, and proportions that follow from them. The science of understanding essences is geometry, which Spinoza, like other seventeenth-century philosophers, holds in the highest esteem. Geometry promises to arrive at true ideas deductively from principles through a guaranteed method. In working through geometrical demonstrations, the mind perceives the truth and gains a model for reasoning. Individual things and their relations follow from the essence of God just as properties and relations follow from the definition of a geometrical figure, and the mind's ability to understand the essence of a figure and what follows from it is paradigmatic of reasoning. This governing analogy of Spinoza's thinking is discussed in Chapter 1 by Valtteri Viljanen, who argues that the example of the circle that contains 'infinitely many [equal] rectangles' (E IIP8S) is the thought-model for Spinoza's ontology. Substance, by definition (E ID3), relates to itself and conceives itself truly, placing *ratio* at the heart of Spinoza's system.

This book is based on the AHRC-funded research project *Equalities of Wellbeing in Philosophy and Architecture*.[1] The project aimed to investigate the relevance of Spinoza's concepts of *ratio* to architecture and housing design. Accordingly, several chapters connect Spinoza and architecture. The architect is one of several exemplars of rational thinking that Spinoza presents us with (alongside, for instance, the 'free man' of *Ethics* Part IV and Moses in the *Theological-Political Treatise*). The architect deduces properties and relations from the essences of geometrical figures in order to design a building. An architect who 'conceives a building in proper fashion' has a true idea of that building (TIE 69) and understands its causes: the architect knows how to build the structure, how it will relate to its human inhabitants, and what good proportions can follow from its essence. The design process can be more 'perfect' as the architect more fully realises the potentialities of geometrical figures: 'ideas are the more perfect as they express a greater degree of perfection of an object. For we do not admire the architect who has designed a chapel as much as one who has designed a splendid temple' (TIE 108; cf. E IVPref.). That design involves the expression of ideas, and is not necessarily a representational process, is the argument of Stefan White in Chapter 9, drawing particularly on Deleuze's 'expressionist' interpretation of Spinoza.

The design of our homes and spaces for work and leisure, and the planning of our urban and rural environments, are key determinants of our wellbeing as individuals and societies. Good architecture provides space for activity and thinking, helps us to avoid harmful causes from the outside world, and provides an environment in which our reasoning and freedom may develop. Bad architecture ignores or works against people's needs for space and shelter and causes obstructions to activity and thinking, negative or harmful interactions, and diminished reasoning and freedom. Following Spinoza's ethics, 'good' and 'bad' here reflect the extent to which architecture affects us positively and aids human flourishing. Gökhan Kodalak, in Chapter 7, considers the potentialities of a Spinozistic 'affective architecture'. Peg Rawes, in Chapter 8, discusses how architecture can aid wellbeing and explores the potential for building on 'good' proportions in housing design and housing policy.

Architecture, for Spinoza, is the art of geometrical thinking, one that can generate positive relations between people, places, and things. *Wall Drawing #122* by Sol LeWitt (1972), a detail of which appears on the cover, does the same thing. LeWitt's wall paintings consist of precise written instructions for painting lines and shapes in certain proportions on a built surface, usually a gallery wall. Teams of people follow these instructions to produce a set of interacting geometrical figures on a vast scale. Based on LeWitt's rationale, the wall paintings realise the potentialities of geometrical essences. However, the paintings do much more than this, for – as Spinoza would have recognised – they are physical beings that are determined not only by their essences but also by their relations to other physical things. LeWitt's instructions cannot capture (and are not intended to capture) the different ways in which each team of painters will interpret and draw them, and how they will interact with each other and their architectural environment. The paintings are expressions not merely of *what geometrical figures can be*, but of *what geometrical figures are* at a specific time and place and under specific conditions of production. The viewer apprehends not an image of eternal essences of circles, triangles, and lines, but a rational and relational working-out of how these essences have been actualised. The viewer follows these spatiotemporally specific 'demonstrations' through his or her own reasoning, just as he or she follows the demonstrations of Spinoza's *Ethics*.

I have chosen one of LeWitt's wall paintings for the cover because it is a physical expression of the complexities of Spinoza's philosophy of *ratio*. Reason and reasons, relations, proportion, geometry, and architectural design are complexly entangled, and this entanglement is present in virtually every aspect of Spinoza's thought. Spinoza's philosophy is a philosophy of *ratio*, as the following chapters reveal.

Note

1. For information, see the Acknowledgements above. Further details are available at www.equalitiesofwellbeing.co.uk.

1

Spinoza's Ontology Geometrically Illustrated: A Reading of *Ethics* IIP8S

Valtteri Viljanen

The *Ethics* is probably the most famous and finest example of a philosophical treatise written in the synthetic geometrical style in which propositions are derived from basic definitions and axioms. This alone leaves little doubt that, for Spinoza, geometry provides the pivotal model for philosophy. However, we should not rush into thinking that *a certain form of exposition* is the only, or even the most important, sense in which geometry informs his philosophy; there are reasons to think that geometricity is ingrained deeper than that in his thought. Most notably, Spinoza's penchant for geometrical examples to illustrate key points of his system signals this. Perhaps the best-known instance of this is the following:

> I think I have shown clearly enough (see P16) that from God's supreme power, *or* infinite nature, infinitely many things in infinitely many modes, i.e., all things, have necessarily flowed, or always follow, by the same necessity and in the same way as from the nature of a triangle it follows, from eternity and to eternity, that its three angles are equal to two right angles. (E IP17S)[1]

In other words, all things as modifications of the single substance follow from the nature (or essence) of that substance, precisely in the way that certain necessary properties follow from the essence of a geometrical figure such as a triangle. This arresting claim is in line with and, I think, the source of Spinoza's no less striking necessitarianism, according to which nothing could have been otherwise since everything takes place with absolute necessity.[2]

In addition to the example concerning (the essence of) a triangle and its properties, I would like to draw attention to three especially prominent illustrations. To take the earliest first, in the *Treatise on the Emendation of the Intellect* Spinoza sets two requirements for a proper definition (of the essence of a finite thing).[3] Here is the first:

> The definition ... will have to include the proximate cause. E.g., according to this law, a circle would have to be defined as follows: it is the figure that is described by any line of which one end is fixed and the other movable. The definition clearly includes the proximate cause. (TIE 96)

In other words, a thing can only be properly defined – and thereby understood – genetically, with geometry showing how this is to be done. The second requirement reads as follows:

> We require a concept, *or* definition, of the thing such that when it is considered alone, without any others conjoined, all the thing's properties can be deduced from it (as may be seen in this definition of the circle). For from it we clearly infer that all the lines drawn from the center to the circumference are equal. (TIE 96)

These two passages are based on the following line of thought: every genuine thing has a definable essence (that constitutes the thing, makes it the thing it is),[4] that essence comes to be generated in a certain way (which is reflected by the first requirement), and from it certain properties necessarily follow (the second requirement latches on to this). All this may – and I think should – be taken to reveal Spinoza's general way of thinking about the fundamental inner structure of all finite things, and, quite characteristically, Spinoza opts for geometrical illustrations to drive his point home.

The second famous example – interestingly in line with the two requirements just mentioned – concerns the way in which our emotions (or affects) have

> certain causes, through which they are understood, and have certain properties, as worthy of our knowledge as the properties of any other thing ... Therefore, I shall treat the nature and powers of the Affects, and the power of the Mind over them, by the same Method by which, in the preceding parts, I treated God and the Mind, and I shall consider human actions and appetites just as if it were a Question of lines, planes, and bodies. (E IIIPref.)

To consider human emotions and actions 'just as if it were a Question of lines, planes, and bodies' may make one think of a mechanistic approach;[5] but since the beginning of the passage refers to causes and properties (which, as we have seen, Spinoza thinks in geometrical terms) and the method is said to be the same as that which Spinoza has applied to God (and thus hardly in any sense 'mechanistic'), it seems evident that he is proclaiming that we are to understand what we feel and do in the same way we understand geometrical objects and their properties.[6]

The third significant geometrical illustration is located in IIP8S of the *Ethics*. In this chapter, I offer an in-depth reading of that illustration and show how it can be used to explicate the whole architecture of Spinoza's system by specifying the way in which all the key structural features of his basic ontology find their analogies in the example. The illustration can also throw light on Spinoza's ontology of finite things and inform us about what is at stake when we form universal ideas. In general, my reading of E IIP8S thus elucidates what it means, for Spinoza, to think geometrically or to consider geometry as a model: fundamentally, geometricity is not a form of exposition but the way in which reality itself is structured.

The Illustration and its Background

As our main geometrical example is designed to illustrate E IIP8 and its corollary, they should be quoted in full. Here is the proposition itself:

> The ideas of singular things, *or* of modes, that do not exist must be comprehended in God's infinite idea in the same way as the formal essences of the singular things, *or* modes, are contained in God's attributes. (E IIP8)

The demonstration is brief in the extreme: 'This proposition is evident from the preceding one [IIP7], but is understood more clearly from the preceding scholium.' E IIP7, in turn, presents what has come to be called Spinoza's parallelism: 'The order and connection of ideas is the same as the order and connection of things.' Discussing this proposition, as difficult as it is central, would take us too far afield; we must simply note that we are indeed in deep Spinozistic waters and move on to the corollary of E IIP8:

> From this it follows that so long as singular things do not exist, except insofar as they are comprehended in God's attributes, their objective being, *or* ideas, do not exist except insofar as God's infinite idea exists. And when singular things are said to exist, not only insofar as they are comprehended in God's attributes, but insofar also as they are said to have duration, their ideas also involve the existence through which they are said to have duration. (E IIP8C)

The proposition and its corollary have been the subject of a prolonged discussion,[7] but I think it is safe to make two main points here, one being ontological, the other epistemological. The ontological point is that formal essences are atemporally contained in their attributes; the epistemological point is that this enables us to have adequate ideas of non-existing things, that is, of things that are not actual at the moment.

The key illustration reads as follows:

> If anyone wishes me to explain this further by an example, I will, of course, not be able to give one which adequately explains what I speak of here, since it is unique. Still I shall try as far as possible to illustrate the matter: the circle is of such a nature that the rectangles formed from the segments of all the straight lines intersecting in it are equal to one another. So in a circle there are contained infinitely many rectangles that are equal to one another. Nevertheless, none of them can be said to exist except insofar as the circle exists, nor also can the idea of any of these rectangles be said to exist except insofar as it is comprehended in the idea of the circle. Now of these infinitely many [rectangles] let two only, viz. [those formed from the segments of lines] D and E, exist. Of course their ideas also exist now, not only insofar as they are only comprehended in the idea of the circle, but also insofar as they involve the existence of those rectangles. By this they are distinguished from the other ideas of the other rectangles. (E IIP8S)

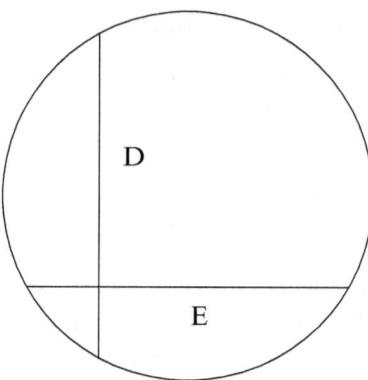

Figure 1.1

The background of the illustration is not difficult to discern: it is in fact proposition 35 of the third part of Euclid's *Elements*:

> *If in a circle two straight lines cut one another, then the rectangle contained by the segments of the one equals the rectangle contained by the segments of the other.*
>
> For in the circle ABCD let the two straight lines AC and BD cut one another at the point E. I say that the rectangle AE by EC equals the rectangle DE by EB. (Euclid 1908: 71)

For our purposes there is no need to go into Euclid's lengthy demonstration of the proposition. It suffices to note that we can derive an infinite number of rectangle pairs by drawing intersecting lines in a circle and that the resulting rectangles are always equal (that is, AE × EC = DE × EB). To connect this to Spinoza's epistemological concerns, there is nothing preventing us having true ideas of the rectangles thus produced, *should they actually exist or not*. This

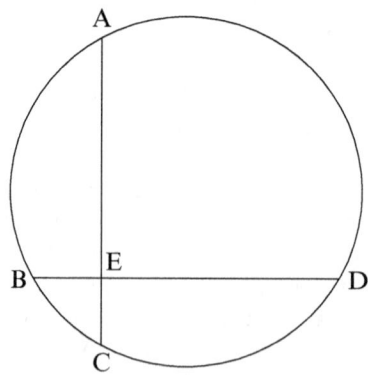

Figure 1.2

reflects an important idea that Spinoza voices in a forceful manner early in his philosophical career:

> If some architect conceives a building in an orderly fashion, then although such a building never existed, and even never will exist, still the thought of it is true, and the thought is the same, whether the building exists or not. (TIE 69)

Spinoza arguably finds Euclid's proposition particularly apt for illustrating that we can form an adequate idea of whatever can be construed 'in an orderly fashion' – for this, the actual (durational) existence of the object is simply irrelevant.

Monistic Ontology and the Geometrical Illustration

The epistemological point is certainly noteworthy; it is, after all, what Spinoza ostensibly wants to convey with the example. But we can go further and show how deeply entrenched the illustration is in Spinoza's system, which bestows upon it explicatory force of a completely different level. The illustration is built on a specific geometric architecture. And that architecture – when given a slight interpretative twist – captures strikingly faithfully *the basic structural features of Spinoza's whole ontology*. I assume that Spinoza himself was at least partly aware of this, for it is, in the end, clear that the illustration contains many central elements of his ontology. But let us consider this again when I have presented my case.

From the Infinite to the Finite

The single substance, the monistic God-or-Nature without which 'nothing can be or be conceived' (E IP15), is of course the most basic element in Spinoza's ontology. We have seen that, for Spinoza (as for so much of the Western philosophical tradition), any genuine thing is endowed with its essence, that which makes it the thing it is. This is no less true of the absolutely infinite substance than it is of finite things. Spinoza calls that which constitutes the essence of substance an *attribute*.[8] A substance can have many attributes (E IP9), each infinite 'in its own kind' (E ID6), but there cannot be an attributeless substance, only, for instance, substance as thinking or substance as extended. In our illustration, *the circle clearly represents an attribute*: after all, the scholium is supposed to help us understand how essences of singular things are contained in their attribute, so the circle as the starting point of the illustration represents the latter.[9] We know only two attributes,[10] thought and extension, and can thus regard the circle as the thinking substance or the extended substance.

What about the lines from which rectangles are formed? Spinoza says that the rectangles are *comprehended* or *contained* in the circle, but it is not immediately clear what this means. The following claim by Charlie Huenemann is, I believe, helpfully correct: '[W]hen X geometrically contains Y, it means that X has sufficient features for producing Y, *in accordance with sanctioned means of construction.*'[11] To give an example of this, recall how Spinoza describes the production of a sphere (in TIE 96). I believe he would approve of the following method of constructing it from extension: first move a point rectilinearly (in extension), then rotate the resulting line by holding its one end fixed and moving the other.[12] In line with this, it is important to note how central *motion* is for the Spinozistic extended substance:

> In examining natural things we strive to investigate first *the things most universal and common to the whole of nature: motion and rest, and their laws and rules*, which nature always observes and through which it continuously acts. From these we proceed gradually to other, less universal things. (TTP ch. 7/G III 102, emphases added)

In other words, extended God-or-Nature produces everything 'less universal' – including rocks, trees, dogs, and human bodies – *through law-obeying motion*. This contention is completely consistent with Spinoza's doctrine of infinite modes, or modes that 'follow from the absolute nature of any of God's attributes' and 'have always had to exist and be infinite, *or* are, through the same attribute, eternal and infinite' (E IP21).[13] In his correspondence, Spinoza tells us that the immediate infinite mode – the mode most directly rooted in its attribute – of extension is *motion and rest* (Ep. 64 to Schuller).[14] And the early *Short Treatise* explains that 'each and every particular thing that comes to exist becomes such through motion and rest' (KV II.Pref.), which all amounts to the claim that extended substance produces particular corporeal things through its infinite immediate mode, namely motion and rest.[15] Thus, I think it is well warranted to say that the immediate infinite mode is the fundamental mode of generation or production within its attribute, and that in the extended substance this mode is lawful motion (and rest).[16]

Given the aforesaid, when Spinoza says that 'the circle is of such a nature' that 'from the segments of all the straight lines' are formed rectangles (E IIP8S), he is illustrating the ontological thesis that particular things of any given substantial attribute (such as extension) come to be produced through its immediate infinite mode (such as lawful motion). We can thus say that *the drawing of the lines* corresponds to *the basic mode of production* within the attribute. To put the point colloquially, it is by 'doing' motion (in the attribute of extension) or thinking (in the attribute of thought) that Spinoza's God-or-Nature produces finite bodies and ideas.

An infinite number of lines can be drawn inside a circle so that in it 'there are contained infinitely many rectangles that are equal to one another' (E IIP8S). We can next focus on an evident element of the illustration that Spinoza does not himself mention: each pair of lines determines or generates *a cutting point*.[17] Here it is helpful to recall how, according to TIE 96, a definition must state the 'proximate cause' that produces the essence; in the illustration, each pair of lines primarily results in a cutting point, which can thus quite naturally be viewed as an essence that constitutes a particular finite thing.[18] The scholium is silent both about cutting points and essences, but the proposition itself is not: 'the formal essences [*essentiae formales*] of the singular things, *or* modes, are contained in God's attributes' (E IIP8). The reference to essences is understandable due to their pivotal role in Spinoza's system; moreover, given that the epistemological point of the proposition concerns the atemporal objects of knowledge, it is completely appropriate to refer more precisely to *formal* essences. Already the early *Metaphysical Thoughts* informs us that 'the formal essence ... depends on the divine essence alone, in which all things are contained. So in this sense we agree with those who say that the essences of things are eternal' (CM I.2). This is not the place for a prolonged discussion of Spinoza's essentialism; it suffices to note that atemporal formal essences are to be contrasted to actual essences and that there are good grounds to consider the former ontologically prior to the latter.[19] Now, Spinozistic particular things are specific, limited ways in which the attribute becomes modified, essences operating as what I would call *attribute modifiers* that constitute those things.[20] Spinoza's geometrical example can help us to grasp the way in which he views the relationship between attributes, immediate infinite modes, and the essences of finite modes: just as line-drawing is the feature of the circle through which are produced specific cutting points, immediate infinite modes (such as motion and rest or infinite intellection) of attributes (such as extension or thought) are the basis for the production of formal essences. These essences, in turn, are manifested differently under different attributes, resulting in such entities as minds and bodies. The so-called Physical Digression of the second part of the *Ethics* indicates that in extension the (formal) essence of a body involves a certain ratio of motion and rest (E IIA2"D); the case of thought is more conjectural, but I would suggest that the formal essence of a mind involves a specific form of affirmation through which objects are conceived.[21]

In addition to the immediate infinite modes, there is one more infinite element in Spinoza's ontology, usually called mediate infinite modes.[22] When asked for an example of them, Spinoza famously gives 'the face of the whole universe' (Ep. 64 to Schuller). This claim is often taken as referring exclusively to extended substance, though it might be intended as attribute-neutral.[23] In any case, I believe the geometrical example can accommodate this idea as well: drawing an infinite number of lines yields an infinite number of cutting points, which can be viewed as an infinite *whole* (of points), just as the face of the whole (mental or physical) universe can be viewed as an infinite whole of formal essences. In this way, the infinite totality of all the cutting points can be used to illustrate the face of the whole physical or mental universe, or mediate infinite modes of their respective attributes.[24] Thereby we have been able to place, proceeding from the infinite to the finite, all the fundamental elements of Spinoza's ontology in one geometrical illustration lifted from Euclid's *Elements*.

The Ontology of Finite Things

We can now begin discussing, in terms of the illustration, the ontology of finite things. Recall the two requirements for a definition of a finite thing (TIE 96): the

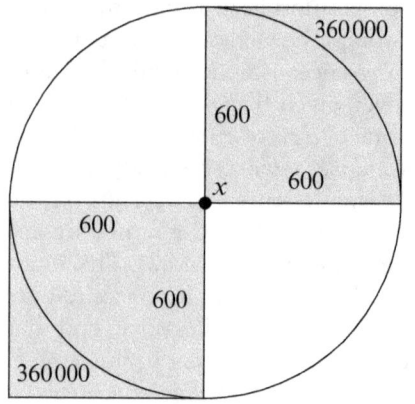
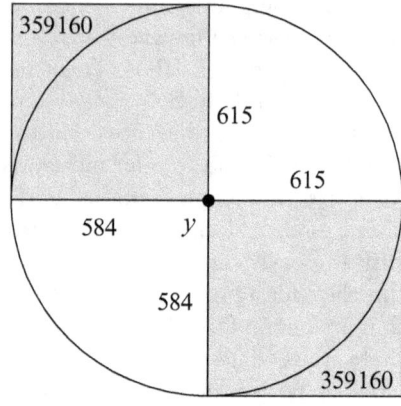

Figure 1.3   Figure 1.4

definition must designate how the thing is to be generated and what necessary properties it has. In the illustration, we can define a cutting point by stating that 1) it is produced by drawing two lines that intersect each other in a circle and that 2) thereby is set a platform from which other specific properties, such as rectangles of equal area, follow. For instance, a point x can be produced by drawing two lines so that two pairs of segments result, each segment having the length of 600 units, thus resulting in equal rectangles the area of which is 360,000 square units.

Let us then draw another point, y, whose segment pairs are both of the length 615 and 584, with the rectangles of 359,160 square units. Points x and y are thus very close to each other but still different in an exactly definable way – the mode of generation has been slightly different, with respectively different segments and rectangles. The circle contains an infinite number of points with their corresponding pairs of segments with different lengths; they form a continuum in which each point is nevertheless individual and expressible in an exact manner. I believe that all this applies to Spinoza's formal essences: they are all individual despite the fact that the basic mode of production is the same in each case (it is only varied in each case), and given that there is an infinite number of them, the points form a continuum in which each individual occupies an exact position, capturable by the individual's proper definition.[25] This illustrates quite accurately what Spinoza says in the *Short Treatise*:

> [A]ll and only the particulars have a cause, not the universals, because they are nothing.
>
> *God, then, is a cause of, and provider for, only particular things.* So if particular things have to agree with another nature, they will not be able to agree with their own, and consequently will not be able to be what they truly are. E.g., if God had created all men like Adam was before the fall, then he would have created only Adam, and not Peter or Paul. But God's true perfection is that *he*

gives all things their essence, from the least to the greatest; or to put it better, he has everything perfect in himself. (KV I.6, emphases added)

This, together with Spinoza's famous contention that 'I say that to the essence of any thing belongs that which, being given, the thing is [NS: also] necessarily posited and which, being taken away, the thing is necessarily [NS: also] taken away; or that without which the thing can neither be nor be conceived, and which can neither be nor be conceived without the thing' (E IID2), gives us good reason to think that finite individuals are endowed with essences unique to their possessors.[26] As depicted above, the illustration can throw light on why and how he thinks like this.

We have seen that things are more than their essences, for certain properties follow from any essence.[27] Properties 'common to all bodies' (E IIP38Dem.) are famously important for Spinoza's epistemological concerns, for what he calls 'reason and the second kind of knowledge' (E IIP40S2) is based on them. This type of property – common to all finite things of a given attribute – can also be expressed in terms of our illustration, for there is a property that follows from *all* the cutting points: rectangles formed from their segment pairs are always equal to each other, be the exact areas of those rectangles what they may. This may be taken to illustrate the way in which, in Spinoza's ontology, all extended things always have a shape and all mental things always involve affirmation (or negation), even though the precise nature of these properties varies immensely, just as do the areas of different rectangles. Taking all the aforesaid together yields what may be called *the full layout of the ontological structure of a finite thing*, which consists of 1) certain causes generating 2) the essence of a thing 3) from which necessarily follow a number of properties. All of these find their analogues in the illustration: 1) drawing of two lines generates 2) a cutting point so that 3) two equal rectangles of a specific area follow.

The illustration can not only help us to understand why and how Spinoza considers finite things to be endowed with their individual essences, but it can also, in an indirect fashion, throw light on the way in which he sees the ontological status of species universals (or species-essences). Now there are, of course, innumerably many ways in which certain points can be demarcated from the totality contained in the circle; Figure 1.5 offers one rudimentary example of such demarcation.

In Spinoza's epistemology, this corresponds to forming universal ideas: we can, for instance, mark off a certain group of essences (of bodies and minds) from the rest of the essences (of physical and mental individuals) based on real features they share (bodies can be, for instance, endowed with a similar ratio of motion and rest, minds with the capacity to reason) and then classify their bearers under a term such as 'human being'. Now this kind of idea of a species is grounded in what really exists, which is arguably why Spinoza is quite happy, despite vehemently criticising imaginatively formed universals (E IIP40S1), to refer to such entities as man, horse, and insect.[28] However, his basic ontology does *not* contain species-essences that would make things what they are – deep down everything

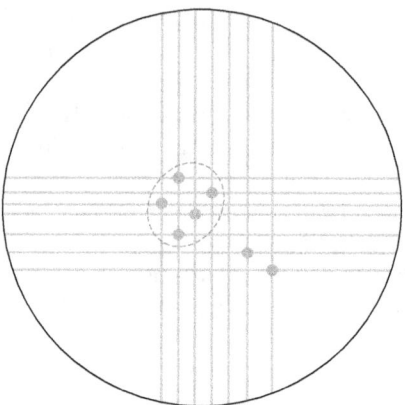

Figure 1.5

finite is, in its (formal) essence, strictly individual. In terms of the illustration, it can be said that the drawing of the lines is uniform in the sense that there is no geometrical reason for classifying different kinds of line-drawings and resulting cutting points (which would correspond to ontologically robust genera and species), even though we can employ different methods to group together points based on, for instance, their position in the circle. Likewise, a universal idea, such as that of the species 'human', is a being of reason (*ens rationis*), something that is 'in our intellect and not in Nature' (KV I.10). But this does not mean that it could not be a highly useful idea, or that it would have no basis in reality.[29]

I have argued above that the illustration can help us to see how attributes and the (immediate) infinite modes are involved in the generation of finite modes; but more is needed to arrive at finite things.[30] What is needed, in addition, is other finite things: Spinoza famously states that 'every singular thing, *or* any thing which is finite and has a determinate existence, can neither exist nor be determined to produce an effect unless it is determined to exist and produce an effect by another cause, which is also finite and has a determinate existence' (E IP28).[31] I would like to finish this section by considering whether there might be a way of finding a place and a role for this idea within the illustration. Stretching it a bit, there is the following option: we could think about the drawing of the lines and the generation of cutting points being determined *from one point to the next* so that each point determines which point is nearest to it in the order of generation; correspondingly, in Spinozistic extension, finite bodies determine the specific form or path that motion and rest take so that particular bodies come to be generated (and affected).[32] However, as the example offers no geometrical rationale for this determination and as it is generally difficult, to say the least, to place features involving duration into something as atemporal as a geometrical figure, I think we should acknowledge that the illustration has here reached its limits. Then again, we are to expect only so much from an illustration; as Spinoza himself notes, 'if anyone wishes me to explain this further by an example, I will, of course, not be able to give one which adequately explains what I speak of here, since it is unique' (E IIP8S). But it does not appear completely impossible to give, within the framework provided by the example, at least the rudiments of an outline of how to understand the way in which finite things are involved in generating other finite things, and in determining each other.

Conclusion

To conclude, I would like to sum up the ways in which the present interpretation of the key geometrical example can enhance our understanding of Spinoza's thought. First, the illustration depicts the ontological priorities involved – and it is of crucial importance that we grasp those priorities, for they structure much of Spinoza's philosophical system. Second, the illustration shows why finite modes are not really but only modally distinct entities in their attributes. Third, it can throw light on how finite essences – which make things what they are – are generated out of such infinite entities as attributes and immediate infinite modes. Fourth, it illustrates how to understand the individuality of essences: they are slightly but nevertheless precisely different from each other. Fifth, it suggests a way in which the mediate infinite mode may be seen as the totality of the formal essences of finite things. Sixth, starting from the core features of reality, it illustrates the generation/essence/property structure of finite things, which structure must be captured by the proper definition of a finite thing. Finally, it suggests why Spinoza thinks about (the ideas of) species as he does.

Let us finish by taking a step back and returning to the question concerning the status of geometry, and the illustration, for Spinoza. All those ontological features find their analogues in the illustration that is, after all, Spinoza's own. Is this just a coincidence? A mere fluke? Or does this tell us something profound about the principles operative behind his system-building? No definite answer can avoid being conjectural. One suggestion would be that the illustration – together with Spinoza's tendency to present key aspects of his system through geometrical examples – in fact reveals a kind of deep structure of Spinoza's thought; that geometry guides his thinking to the extent that he might well have lifted another example from Euclid's *Elements*, and the one he ended up choosing merely happened to suit his purposes particularly well. However, many would no doubt find this too bold and object to giving too much weight to observations concerning a single example. Whatever one's stand on the issue happens to be, at least this much is certain: the evidence is there, and of the type of clarity Spinoza himself celebrates.[33]

Notes

1. I am using Curley's translations.
2. I will here set aside the much-debated question whether Spinoza is a necessitarian or merely a determinist. However, a number of contributions since Garrett (1991) have

shown that Spinoza endorses necessitarianism; for a recent discussion of the topic, see Jarrett (2009).
3. It is important to note that, as E IP8S2 states, 'the true definition of each thing neither involves nor expresses anything except the nature of the thing defined'. In other words, Spinozistic definitions concern precisely *essences* or *natures* of things.
4. Cf. especially: 'Understand the definite nature, by which the thing is what it is, and which cannot in any way be taken from it without destroying it, as it belongs to the essence of a mountain to have a valley, or the essence of a mountain is that it has a valley' (KV I.1).
5. See, for example, Marshall (2013: 3).
6. For more on this, see Viljanen (2011: 149–57).
7. See, for example, Gueroult (1974: 100–2); Jarrett (1990: 162); Koistinen (1998: 74–5).
8. E ID4 states that 'by attribute I understand what the intellect perceives of a substance, as constituting its essence', which has given rise to a perennial debate as to whether attributes are only subjective ways of perceiving the substance (see esp. Wolfson 1961 [1934] I: 142–57) or are objectively constitutive of the substance (see esp. Gueroult 1968: 428–61). Fortunately, this issue has no bearing on the present interpretation and can thus be left aside.
9. Given the nature of the illustration, it is certainly easier to think of extended than thinking substance here, but Spinoza's point is, of course, meant to be attribute-neutral.
10. Again, I am here leaving aside a traditional topic of discussion and just take Spinoza at his word when he clearly implies in Ep. 56 (to Boxel) that there are more attributes than the two we are acquainted with.
11. Huenemann (1999: 233, emphasis added). See also Marshall (2008: 74 n.54).
12. For the production of a line, see TIE 108.
13. Spinoza's doctrine of infinite modes is notoriously difficult; for a recent helpful discussion, see Melamed (2013: ch. 4).
14. Correspondingly, the immediate mode of thought is the 'absolutely infinite intellect' (Ep. 64 to Schuller). For discussion, see Melamed (2013: 134 n.54).
15. However, as Yitzhak Melamed (2013: 116) points out, according to the *Ethics* (but obviously not the *Short Treatise*), only infinite modes can follow from infinite modes. As a consequence, given that the *Ethics* is the authoritative work here, something in addition to infinite modes must be operative when finite modes are produced through an infinite mode; I discuss this topic below.
16. Spinoza famously treats motion together with rest, talking often about motion and rest as a package in which 'rest is certainly not Nothing' (KV II.21). This issue has no bearing on the present interpretation, so I will not attempt to analyse it further; for some discussion, see Melamed (2013: 135).
17. Here, of course, we have to assume that lines are drawn only horizontally and vertically, otherwise a point could be drawn in infinite different ways, and so the point could be defined in infinitely many different ways.
18. As will become clear in what follows, I regard essences of finite things as individual, unique to their possessors; as the useful survey by Martin (2008) shows, this is the dominant view in the literature. See also Hübner (2016).
19. I argue for these (not particularly contentious) claims especially in Viljanen (2011: 23–4; 2014: 264–5). See also Garrett (2009: 285–6).
20. See especially Viljanen (2011: 75).

21. E IIP49S states that we are not to consider ideas 'as mute pictures on a panel' because each idea 'involves an affirmation or negation'. This reveals, I think, what is essential to the mind as it actively forms ideas 'because it is a thinking thing' (E IID3). Even though Spinoza says nothing about this, for systematic reasons the ideas resulting from the mind's essential affirmative activity can be viewed to stand in a certain interrelationship with each other just as the parts of our bodies 'communicate their motions to each other in a certain fixed manner' (E IIA2″D).
22. See E IPP22–23. For a discussion of terminology concerning the infinite modes, see Melamed (2013: 114–16).
23. For discussion, see Melamed (2013: 134 n.54, 136).
24. This being said, I agree with Melamed (2013: 136) that the issue of mediate infinite modes is 'pretty foggy'. Moreover, my interpretation is not far removed from Melamed's (2013: 136) assessment concerning extension: 'It thus seems quite plausible that the face of the whole universe is indeed this infinite individual that contains all bodies as its parts.' In other words, whereas Melamed sees the face of the whole (extended) universe as containing all bodies, I see it as containing all the formal essences of bodies.
25. For the purposes of the illustration we can disregard the fact that the sectors of the circle mirror each other in a way that produces identical segment pairs.
26. I examine the connection between the *Short Treatise* and E IID2 in more detail in Viljanen (2015: 186–7). For a very helpful recent account of E IID2 and individual (or particular) essences, see Hübner (2016: 64–5, 68–9). I discuss below what are usually called species-essences.
27. See Ep. 83 to Tschirnhaus for Spinoza's discussion of whether there are things from whose definition only one property can be derived.
28. 'For example, a horse is destroyed as much if it is changed into a man as if it is changed into an insect' (E IVPref.).
29. This view of universal ideas and species-essences squares very well with, and lends some additional support to, the recent account presented by Karolina Hübner. She sums up her interpretation as follows: '(i) only particulars and their essences have formal reality; (ii) the essences of such actually-existing particulars are unique; however (iii) Spinoza's metaphysics also allows for more general species-essences; (iv) such species-essences are only insofar as they constitute ways that finite minds spontaneously think of certain genuinely similar particulars as the same in some respect, when they abstract and compare their properties' (Hübner 2016: 74). To my mind, this is basically right, and in fact considerably less speculative than Hübner (2016: 80) herself thinks.
30. See note 15 above.
31. As E IP11S states, finite things 'come to be from external causes'; for discussion, see Viljanen (2011: 71).
32. The order of generation is temporal in character, so how could temporal features be related to the illustration? The determinations between points would (somehow) need to be convertible into determinations of time and place, through which durational existence – a causal process unfolding in time – could come to be formed (see also Marshall 2008: 74; Viljanen 2014: 268). This would mean that interdeterminations of formal essences determine the nature (time and place, and thus duration) of actual existence. However, not even the third kind of (intuitive) knowledge captures this; as the young Spinoza states, 'it would be impossible for human weakness to grasp the series of singular, changeable things' (TIE 100).

33. I would like to thank Karolina Hübner, Beth Lord, Juhani Pietarinen, Arto Repo, Justin Steinberg, and Xiaohui Zhang for very helpful comments on earlier versions of this chapter. I am also grateful to the audiences at Aberdeen, HU Berlin, Jyväskylä, and Turku. Finally, I would like to acknowledge that the work on this chapter has been financially supported by the Academy of Finland (project number 275583).

2
Reason and Body in Spinoza's Metaphysics

Michael LeBuffe

Traditional labels for central tenets of Spinoza's metaphysics support the view that, in response to Descartes, Spinoza maintains a strict symmetry between thought and extension. Descartes had argued that existence includes some things that are minds, others that are bodies, and still others that are both minds and bodies. Thus God and angels are minds; a chair is a body; and I, in at least one important sense, am both. In what has been known as the 'dual aspect theory', Spinoza rejects this view. He contends instead that all things, including God and ordinary objects and all persons, are in one way thinking and in another way extended. In the case of a person, for example, I am a mind; I am a body; and my mind and my body, although they are of different attributes, are identical. In his accounts of causal relations, Descartes had argued that mind can interact with mind in isolation from all body; that body can interact with body in isolation from all mind; that minds can affect bodies; and that bodies can affect minds. The causal history of the world for Descartes, then, might include chapters that are wholly mental, chapters that are wholly corporeal, and chapters that include interaction between mind and body. In what has been known as his 'parallelism', Spinoza rejects all mind–body interaction and maintains, moreover, that the causal interactions of bodies are in some sense the same as the causal interactions of minds and vice-versa. Bodies do affect other bodies and minds affect other minds. Mind and body are parallel in the sense that the order of those causal interactions is the same: there is only causal history of the world.

The dual aspect theory and parallelism undoubtedly offer advantages over their Cartesian rivals, notably in the accounts that they offer of the mind–body relation in human beings. Spinoza's views raise problems of their own, however. A particularly pressing problem concerns causation among finite things. Ordinarily we might suppose, whenever A causes C and B is identical to A, that B is also the cause of C. For example, if Lorde has inspired the youth of the world, and if Lorde is identical to Yelich-O'Connor, then it just seems clear that Yelich-O'Connor has inspired the youth of the world. Spinoza cannot admit that suppositions of this sort are always warranted. In particular, while the dual aspect theory makes bodies and minds identical, parallelism rules out mind–body interaction. If a corporeal cause, C_B, has a given effect, E_B; and the cause is identical with something ideal, C_M; then, given ordinary assumptions, we would have to conclude that C_M

also has the effect, E_B. For a mental cause to have a corporeal effect, however, is a violation of parallelism.

Like some other commentators, I think that Spinoza's response to this problem is to reject the ordinary supposition that anything identical to a given cause has the same effects as that cause.[1] Spinoza does insist that every corporeal cause is identical to some mental cause. Despite this identity he also holds that only corporeal causes have effects on bodies and that only mental causes have effects on minds.

This chapter starts from a conviction of mine, which I will not defend here in its full generality, that Spinoza holds a similar conception of reason.[2] That is, I think that on the account of the *Ethics* only a reason that is itself ideal explains something ideal and only a reason that is itself corporeal explains something corporeal. Just as Spinoza would find it to be a mistake to take something thoughtful to cause something extended, so he would find it to be a mistake to take something thoughtful to be a reason for something extended.[3]

Although Spinoza's close association of reasons and causes in the *Ethics* lends the view some initial appeal, this conception of reason may strike some readers as odd or even as a kind of category mistake. For, beyond suggesting that reasons, like causes, are restricted to particular attributes, it suggests that reasons might be corporeal. Indeed, because Spinoza endorses a version of the principle of sufficient reason (hereafter, PSR), on which there must be a reason for whatever exists, and because he takes bodies to exist, the view that reasons for body must be themselves corporeal positively requires that many reasons be corporeal reasons. The notion of a corporeal reason might seem just incoherent, however, if one thinks that a reason is not the sort of thing that can be extended.

'Reason' in English, like its closest Latin counterpart *ratio*, has a broad range of meanings, and the reflexive response on which reasons just cannot be corporeal may simply result from an emphasis on some rather than other senses of the term. Certainly a reason that someone has consciously in mind, as a basis for some action, seems as though it has primarily a psychological sense. So does the notion of reason as a process of thinking, for example, from some premises to some conclusions. The sense of reason most relevant here is that of explanation, and for this sense, I think that there is a natural and clear way to understand reasons, particularly for things that are themselves corporeal, to be corporeal. Suppose, for example, that a disc slides across the ice. What is the reason, or explanation, for this? The ice is slick. Such a reason, even if understanding it would be a mental act, seems to be corporeal.

Whatever our own intuitions about corporeal reasons, I accept these implications of parallelism for the interpretation of Spinoza together with the burden of argument that accompanies them. The thesis of this chapter, however, is still more narrow: finite bodies, on Spinoza's account, are defeasibly and finitely self-explanatory. That is, they supply their own reasons, and, of course, because they are bodies, those reasons are corporeal. The Latin term *ratio* has several different meanings, including both that of the English term 'ratio' and many of the meanings of the English term 'reason'. Where Spinoza writes about a particular body's *ratio*, I think that he is best understood to refer both to a characteristic ratio or proportion of motions among that body's parts and also to the reason that defeasibly and finitely explains the body's existence. What I mean by 'defeasibly' is that the reason a body supplies for its own existence may not in fact explain the body's existence. If it does not, then like a cause that would have had a given effect but for influence of other causes, what it explains will not come about: the body will not exist. What I mean by 'finitely' is that, while each body does explain its own existence, it is not a complete explanation. The slickness of the ice is a finite explanation of the movement of the disc, in this sense, because other things are also important to the explanation.

The argument will have the following structure. I will start with an account of the PSR in the *Ethics*. This will contribute to the case for my thesis. I want to show that Spinoza does not in his most detailed account of the PSR invoke a conception of reason that we frequently associate with sufficient reason and that is distinctively ideal: he does not invoke God's will. Moreover, even at this high level of generality (an account of reason which applies to all existences and to all attributes), Spinoza seems to take reasons important to human beings to be corporeal. My discussion of the PSR will have another point as well, however, that will inform the rest of my discussion. It will show that Spinoza conceives of reasons as being either internal or external to existing things. That distinction, then, will govern my discussion of particular reasons for body in the *Ethics*. I will show that both internal reasons and external reasons for body are, as Spinoza understands them, corporeal. The internal reason for a given body, it will turn out, is the reason or *ratio* that is that body's nature.

The Demand for Reasons

On any familiar version of the PSR, existence and changes to existence must have a sufficient reason. I hope to show that, on the argument of the *Ethics*, such reasons can be – indeed the most important reasons to human beings are – corporeal. However, that project faces an immediate obstacle: a reason may not seem to be, as it is ordinarily understood, the sort of thing that could be corporeal. Instead it would seem to be a proposition, a concept, an idea, or a motive. In this section, I will offer an interpretation of the PSR as Spinoza presents it in the *Ethics*. Although I will offer some argument for the conclusion that, even in his broadest account of reason, Spinoza considers some familiar sorts of reasons to be corporeal, my case is largely negative: I will argue that the demand for reason in the *Ethics* does not characterise reason in psychological terms. Spinoza's account of sufficient reason does not itself present reasons as propositions, concepts, ideas, or motives. Notably, Spinoza's version does not include the common psychological sense of reason most closely associated with the explanation of particular existences from the PSR, on which reason is a kind of motive. So there is no basis in Spinoza for taking reason to be ideal beyond our own ordinary understanding of reason, a risky tool to use as a guide to the meaning of technical terms in the *Ethics*.

A comparison of well-known passages from Spinoza and Leibniz shows that Spinoza's presentation of sufficient reason in the *Ethics* lacks its most familiar psychological sense. The work of interest, Leibniz's 'Principles of Nature and Grace', was written in 1714, decades after Spinoza's death. I do not mean to suggest in discussing it that Spinoza responds to Leibniz (although undoubtedly Leibniz does respond to some extent to Spinoza).[4] The point of the comparison, rather, is to compare a well-developed and familiar psychological account of sufficient reason to what I take to be a very different account in Spinoza's *Ethics*.

Leibniz begins his discussion of the PSR by distancing the search for sufficient reasons from the level of physical inquiry, suggesting already that he will not find any reasons in bodies:

> §7. So far we have spoken only at the level of *physical* inquiry; now we must move up to the *metaphysical*, by making use of the *great principle*, not very widely used, which says that *nothing comes about without a sufficient reason*; that is, that nothing happens without its being possible for someone who understands things well enough to provide a reason sufficient to determine why it is as it is and not otherwise. (Leibniz 1998: 262)

Leibniz introduces a principle here on which whatever happens happens for a reason, and that reason can be, in principle, fully known. He takes the account of such reasons to be metaphysical rather than physical or, perhaps, merely physical.

In the next section Leibniz explains why such a thing could not be a body. He offers two slightly different reasons, one from the nature of body and one from the order of causes in the deterministic chain of motions:

> §8. Now, the sufficient reason for the existence of the universe can never be found in the series of contingent things, in bodies and their representations in souls, that is; because matter in itself is indifferent to motion or rest, and to this motion or that. Therefore we could never find in matter a reason for motion, and still less for any particular motion. And since any motion which is matter at present comes from previous motion, and that too from a previous one, we are no further forward if we go on and on as far as we like; the same question will still remain. Therefore the sufficient reason, which has no need of any further reason, must lie outside that series of contingent things, and must be found in a substance which is the cause of the series: it must be a necessary being, which carries the reason for its existence within itself, otherwise we still would not have a sufficient reason at which we can stop. And that final reason for things is what we call God. (Leibniz 1998: 262)

The first reason Leibniz offers for the conclusion that sufficient reason could not be body depends upon an understanding of the nature of body that makes it 'indifferent to motion or rest, and to this motion or that'. Suppose that a billiard ball is rolling across the table, and we want to understand why it is rolling in just that way. Leibniz argues that by its nature, it could be rolling any number of different ways or even not rolling at all, so we will not a find a reason in its nature. The second reason that Leibniz offers responds to a different strategy for finding corporeal reasons: why might we not look to some other body for an explanation of a given body's motion? Leibniz responds to this suggestion that such a reason could never be sufficient; that is, it could never fully explain the motion because we would have to continue backward infinitely in the chain of particular motions, searching for further reasons. So neither the nature of the particular body in question nor reference to other bodies will explain a body's motion. We must, Leibniz argues, therefore look for a final reason in God.

God, Leibniz makes clear at §8, is Himself self-explanatory, so any chain of reasons will stop with God. Leibniz proceeds to give an account of the sufficient reason that God provides for the whole system of created things, on which it not a physical reason at all. Instead, it is an inherently psychological kind of reason, a choice:

> §10. It follows from the supreme perfection of God that in producing the universe he chose the best possible design, in which there was the greatest variety, together with the greatest order. (Leibniz 1998: 263)

To take stock, Leibniz finds nothing in the nature of a particular body to explain its particular motion; he also finds nothing in bodies external to the particular body; but he does find a reason in God's choice of the best. The complete explanation will be psychological.

Let us turn now to Spinoza's invocation of the PSR in his argument for the existence of God at E IP11Dem.2:

> For each thing there must be assigned a cause, *or* reason, as much for its existence as for its nonexistence. For example, if a triangle exists, there must be a reason *or* cause why it exists; but if it does not exist, there must also be a reason *or* cause which prevents it from existing, *or*, which takes its existence away.
>
> But this reason, *or* cause, must either be contained in the nature of the thing, or be outside it. E.g., the very nature of a square circle indicates the reason why it does not exist, viz. because it involves a contradiction. On the other hand, the reason why a substance exists also follows from its nature alone, because it involves existence (see P7). But the reason why a circle or triangle exists, or why it does not exist, does not follow from the nature of these things, but from the order of the whole of corporeal Nature. For from the [order] it must follow either that the triangle necessarily exists now or that it is impossible for it to exist now.
>
> These things are evident through themselves, but from them it follows that a thing necessarily exists if there is no reason or cause which prevents it from existing. Therefore, if there is no reason or cause which prevents God from existing, or which takes his existence away, it must certainly be inferred that he necessarily exists.[5]

Spinoza draws a distinction between two classes of things to be explained that is similar to the one that we see in Leibniz: what we ordinarily regard as eternal things belong to one class; what we ordinarily regard as contingent things belong to another. Moreover, the sort of reason that Spinoza supplies for the first class is also similar to that supplied by Leibniz. Leibniz's God is self-explanatory. Spinoza writes here that substance is self-explanatory. There is only one substance for Spinoza, however: God. Clearly, he also takes God to be self-explanatory.

Turning to those existences that are of interest to us, a first point to make is that Spinoza could not offer a reason for them anything like the one that Leibniz offers, even if it were psychological. Leibniz explicitly makes such reasons depend upon a kind of personhood in God: it is God's choice of the best that supplies the reason for particular motions. Spinoza denies the sort of personality to God that Leibniz depends upon, regarding such a view as a kind of misguided anthropomorphism, and he famously rejects teleology in nature.[6]

It is clear by the end of the passage, I think, that Spinoza takes reasons of the sort that interest us to be corporeal. In considering the existence of extended things that we would ordinarily consider to be contingent – a particular circle or a particular triangle – Spinoza writes that it is the 'order of the whole of corporeal Nature' that provides the reason external to them for their existence or non-existence. Insofar as he characterises such reasons at all, then, Spinoza takes them to be corporeal.[7]

Here we arrive at a clear disagreement between Spinoza and Leibniz, which rests, I think, in a disagreement about necessitarianism. Leibniz is clearly a determinist: he thinks that one could look outside of a given body for a cause of its motion. He rejects, however, any view on which knowledge of such causes could yield a reason sufficient for the motion. His position rests, it seems, on the conviction that even knowledge of the whole order of events could not explain the motion because one could not, from this knowledge, explain why that order of events rather than some other order exists. In the end, it is God's choice of the best that explains why this order of events, including this particular motion, exists. Of the various orders of events that are possible, God chooses the best one. Spinoza, by contrast, takes the order of all corporeal nature to explain fully any particular corporeal existent. Knowledge of the whole order of events could be a sufficient reason for Spinoza just because, on Spinoza's view, there turns out to be only one possible order of events.

The passage from E IP11Dem.2, then, yields a clear difference from Leibniz in Spinoza's conviction that some reasons are corporeal. It also yields a clear response to one of Leibniz's objections to the project of finding reasons in bodies. Leibniz's second reason, from §7, was that we cannot look to external causes of particular motions for sufficient reasons. Spinoza clearly disagrees. He argues that the reasons for all contingent things are, in fact, external to them.

This point, that Spinoza takes external reasons to explain bodies, may suggest that Spinoza would accept the first point that Leibniz makes about body. One might think, after all, that if reasons are external to a thing, then they are not internal. Spinoza writes, notably, at the beginning of the second paragraph of the quoted passage, that 'this reason, *or* cause, must either be contained in the nature of the thing, or be outside it'. The sentence suggests an exclusive disjunction: if the reason is external to a thing, then it cannot be contained in the nature of the thing. So understood, we have some evidence to conclude that corporeal reason for a given body is never in the body itself for Spinoza. Perhaps, as Leibniz believes that the nature of body is indifferent to motion, so Spinoza believes that the nature of body is indifferent to existence.

On the other hand, a given body is itself part of the order of the whole of corporeal nature. If that order is the reason for the body, then the body is part, perhaps a small part, of the reason for its own existence. On such an interpretation, we might emphasise the phrase 'contained in' in the crucial disjunction. Suppose that the reason for a finite body is body but is not fully contained in the body. Rather it is fully contained in something more than the body. If this interpretation were correct, then Spinoza would not after all have to hold a view like Leibniz's. He could hold that the nature of a body is not indifferent to its own existence. Adjudicating this issue, then, is the central question moving forward: for Spinoza, does a particular thing give a reason for itself?

Body and Particular Reasons

It may seem odd to think about a thing's nature contributing to the explanation of its own existence. Perhaps it is particularly so if one identifies reasons with causes. I think of efficient causes, at least typically, as preceding their effects. Such a view might require, then, the nature of a thing to exist somehow before the thing itself so that it can do the necessary metaphysical work of pushing the thing into being.

On reflection, however, I think that it is the correct position to say that the nature of a thing contributes to the explanation of its existence. Perhaps seeing this point depends upon taking the identification of causes and reasons in Spinoza to be more a reduction of causes to reasons than a reduction of reasons to causes. Crows exist, at least in part, because the sort of thing that they are can exist, given the various external constraints on them. They exist because they stick together, eat a lot of different things, and possess a lively intelligence. On the other hand, dementors – the soul-sucking, darkness-lurking creatures of the Harry Potter books – do not exist at least in part because, although there is plenty of darkness, there are no souls to suck. Dementors are just not the sort of thing that could exist. Perhaps a similar point could be made in response to Leibniz. A billiard ball does not have a nature such that it is a sufficient reason, that is, a complete explanation, for its motion. Nevertheless, the nature of a billiard ball – its hardness, its roundness, and so on – surely contributes to any explanation of its motion. An anvil would not respond to external causes in the same way.

In this section, I will argue that Spinoza does take the reason for a given finite body's existence to be internal to it. Such a reason is not in itself sufficient for existence, of course. Only God is fully self-explanatory. The finite thing itself is, however, part of its reason. I think that Spinoza is very clear about this. He takes the nature of any finite thing to be a defeasible tendency to cause its own existence. For bodies, Spinoza calls the reason explaining any effect of a given body the *ratio* of motion and rest among the body's parts.

Spinoza's definition of 'individual' in the physical discursus following E IIP13 is the principal evidence for this interpretation in the *Ethics*:

> Definition: *When a number of bodies, whether of the same or of different size, are so constrained by other bodies that they lie upon one another, or if they so move, whether with the same degree or different degrees of speed, that they communicate their motions to each other in a certain fixed manner* [certa quadam ratione], *we shall say that those bodies are united with one another and that they all together compose one body or individual, which is distinguished from the others by this union of bodies.*[8]

Curley translates *ratio* here as 'manner'. This is correct. 'Ratio' or 'proportion', which Curley uses four times in similar contexts in the demonstration and scholium to E IVP39, are also correct. The term has a variety of connotations. I do think, however, that Spinoza capitalises on the different meanings of the Latin term in order to find what he needs: a distinctively corporeal kind of reason.

One might well think, initially, that it is merely a coincidence that Spinoza uses a single term to describe two different, important concepts in the argument of the *Ethics*. I will build a case for taking ratios to be, at the same time, reasons by trying to make it more and more difficult to maintain this opinion. Whatever work reasons do to explain existence in the *Ethics*, that is the work that ratios do with respect to body.

As we have seen, one of the few things that Spinoza writes about reason in his characterisation of the PSR is that a thing whose existence is not explained by something external to it has a reason internal to it in virtue of which it exists. A first point to make, then, is that the ratio characteristic of an individual body, which Spinoza describes in his definition of an individual, is necessary for the individual's existence.

In early propositions of *Ethics* III, Spinoza argues that nothing can be destroyed except through an external cause (E IIIP4); that each thing, to the extent that it can by its own power, strives to persevere in being (E IIIP6); that this striving is the actual essence of the thing (E IIIP7); and that those things that increase or diminish the power of the body likewise, considered as ideas, increase or diminish the power of the mind (E IIIP9). These propositions establish the basis for Spinoza's theory of the affects, on which an individual's striving is its desire (E IIIP9S) and increases or decreases to the power of striving are its passions (E IIIP11 and E IIIP11S). The argument of these propositions is somewhat self-contained because Spinoza takes E IIIP4 to be self-evident. This insularity makes it difficult to establish a connection between it and earlier doctrines of the *Ethics*. Nevertheless there are two ways of connecting the doctrine of striving and the account of an individual body from *Ethics* II.

I will start with the more complex but also more informative way. Lemmata

5 and 7 of the physical discursus following E IIP13 refer back to the definition of the individual in describing ways in which an individual can retain its nature despite change:

> E IIL5: If the parts composing an Individual become greater or less, but in such a proportion that they all keep the same ratio of motion and rest to each other as before, then the individual will likewise retain its nature, as before, without any change of form.
> E IIL7: Furthermore, the individual so composed retains its nature, whether it, as a whole, moves or is at rest, or whether it moves in this or that direction, as long as each part retains its motion, and communicates it, as before, to the others.

Both lemmata clearly invoke the definition of the individual, which Spinoza cites in both demonstrations. Lemma 5 associates the *ratio* closely with the individual's nature or form. Spinoza argues there that it does not matter if the parts composing an individual grow or shrink; it is the same individual so long as the ratio remains the same. Likewise, Spinoza argues at L7 that so long as the ratio remains the same, the motion or rest of the whole will not change an individual's nature. Both lemmata, then, associate the ratio characteristic of an individual body with its nature or form. Indeed, later in the *Ethics*, at E IVP39Dem., Spinoza refers to his definition of an individual as an account of form: 'what constitutes the form of the human body consists in this, that its parts communicate their motions to one another in a certain fixed proportion [*ratio*] (by the definition)'. Taking nature and form to be equivalent in these contexts, this, then, is a first important result.

1. An individual body's ratio is its nature [by E IIL5 & 7, E IVP39Dem.]

The lemmata offer assurances that we human beings might take to hold special importance, since human individuals grow, shrink, and move. In a series of postulates following E IIL7, which describe the human body, Spinoza offers more detailed information. Postulate 1 will be of particular importance here:

> E IIPost.1: The human body is composed of a great many individuals of different natures, each of which is highly composite.

Spinoza does not offer arguments for Postulate 1 or for any of these postulates. Given their place immediately following the lemmata, however, we might assume that he takes the lemmata to hold special relevance for them. That is, at Postulate 1 he means to emphasise the point that a human body, despite its highly composite nature, need not lose its nature if its parts grow or shrink or if it moves.

A different postulate, from the beginning of *Ethics* III, vindicates this assumption by gathering many of the relevant passages:

> E IIIPost.1: The human body can be affected in many ways in which its power of acting is increased or diminished, and also in others which render its power of acting neither greater nor less.
>
> This postulate, or axiom, rests on Post.1, L5, and L7 (after IIP13).

This postulate is the clearest connection in the argument of the *Ethics* between the notion of ratio in the physical discursus and the notion of a striving to persevere in *Ethics* III. It also presents a gap in Spinoza's argument, however, and an ambiguity that is difficult to resolve. The gap is in the transition from the nature or form of an individual, which characterises individual bodies in the physical discursus, to the power of acting, which characterises the human body at E IIIPost.1. Clearly the terms capture something similar. Postulate 1 and the lemmata of the discursus concern circumstances in which the body is affected and form is not. Postulate 1 of *Ethics* III makes a general claim about circumstances in which the human body is affected and its power of acting does not, or does, change. Power of acting, which Spinoza uses at E IIID3 in defining 'affect' but which does not appear in the discursus, is a more complex notion than that of form or nature, however. What remains ambiguous is whether Spinoza identifies nature or form with just one aspect of a body's power of acting – the body's activity that might then be qualified as more or less powerful – or whether he identifies it with power of acting as a whole. In the former case, we would interpret E IIIPost.1 as a claim that the body's nature or form is something that the body retains with more or less power, such that the power of a body can change while its nature remains the same. In the latter case, we would interpret the postulate as something closer to a restatement of the lemmata of the physical discursus: a claim that some effects on the body change its nature but that others do not.

Neither option is absolutely warranted by the argument of the *Ethics*, and the problem of whether the nature of an individual human being is its activity or its power of acting itself arises again in different contexts. Nevertheless, it seems best to take the former option, on which the notion of power of action introduces something in addition to the lemmata which make claims only about action. On this view, then, the form or nature of the human body is its activity, which it can then possess with greater or less power. A passage from the end of the Preface to *Ethics* IV provides some evidence for this interpretation:

> When I say that someone passes from a lesser to a greater perfection, and the opposite, I do not understand that he is changed from one essence, *or* form, to another. For example, a horse is destroyed as much if it is changed into a man as if it is changed into an insect. Rather, we conceive that his power of acting, insofar as it is understood through his nature, is increased or diminished.

Here, Spinoza argues that changes to perfection or to power of acting can take place while the form of a body does not itself change. That suggests that power of acting is not identical to form, or, what is the same in these contexts, nature or essence.[9] On the basis of passages such as this one, then, I will suggest that it is best to find at E IIIPost.1 an identification of an individual body's nature or form with that body's activity.

2. An individual body's nature is its activity [by E IIIPost.1]

So understood, the postulate describes two sorts of changes: those that affect the power with which the body acts and those that do not. A body will certainly be destroyed if it loses all power of activity. So a decrease in the power with which it acts is a serious harm to it. Such a change, however, is not itself a change to the activity of the body. It is merely a change to its power. So it does not, unless it eliminates all power, destroy the body. Perhaps, since the activity of the body is a ratio of motions of its parts, power is something like the ability of the body to maintain this ratio in hostile environments. If this is right, then we will say that a more powerful body maintains its activity in a greater range of hostile environments or in more hostile environments. A less powerful body maintains its activity in a smaller range of hostile environments or in less hostile environments. A body with no power at all maintains its activity in no environments: it ceases to exist. A body with the greatest power would maintain its power in all environments.[10]

What is new with the theory of affects in *Ethics* III, then, is a detailed account of the power of finite individuals, which Spinoza associates in individual bodies with their variable ability to maintain their characteristic ratio and in all things with their perfection and reality.[11] While their application to individuals is new in *Ethics* III, however, these notions do appear together in Spinoza's account of substance in *Ethics* I. Attention to Spinoza's treatment of them there can show the importance of an individual body's ratio to its existence. I think that it shows that ratio plays an explanatory role for the existence of an individual body similar to that which the reason internal to substance plays in explaining the existence of substance.

In the third demonstration to his claim that God necessarily exists at E IP11, Spinoza invokes the notions of power, existence, and perfection. He writes, first, that not to be able to exist is to lack power and to be able to exist is to have power. Next, at a scholium following the demonstration, he writes that perfection asserts the existence of a thing and that imperfection takes it away. Thus the three notions are most clearly related in the account of God: the most perfect, most powerful thing, exists most clearly, or, as Spinoza writes, absolutely (E IP11S). The argument, which is a kind of cosmological argument for God's existence, proceeds from the observation that we, finite things, exist: if we, beings of limited power, exist, but God, the most powerful being, does not exist, then, since to be able to exist is to have power, we would be, absurdly, more powerful than what is most powerful.

Ethics III builds upon this account of finite existents. God, by nature, is absolutely active and so absolutely exists. An individual body's nature is also to exist, but, Spinoza suggests at the beginning of *Ethics* III, it can exist with more or less power, and, if it loses all power (or perfection), it can cease existing altogether.

The central propositions related to the striving to persevere in being reaffirm the association between activity and nature, and they characterise the effects of that activity: perseverance in being. At E IIIP6, Spinoza characterises particular things as expressions of God's own power and argues that each thing strives by its own power to persevere in its being. Power is for bodies the power of activity that Spinoza invokes so suddenly at E IIIPost.1: Spinoza refers to power as the power by which a thing is and acts in the demonstration to E IIIP6. Because activity is form or nature, the proposition implies that, for a body, the causal tendency of the ratio of motion and rest is being.

3. An individual body's activity produces its existence[12] [by E IIIP6]

A comparison of the use made of the notions of power, perfection, and existence in Spinoza's argument for the existence of God at E IP11 and of the use made of the same notions in Spinoza's accounts of individuals at the beginning of *Ethics* III suggests that the reason for an individual's existence will be, properly re-described of course, similar to the reason for God's existence. If God exists absolutely and from the nature of substance, then we ought to conclude that individual things exist something less than absolutely and from their nature as individual things. That is, the reason for the existence of individual bodies is precisely what Spinoza calls a reason, the characteristic ratio of the motions of their parts. This same inference is warranted by the results that I have summarised in this discussion:

1. An individual body's ratio is its nature [by E IIL5 & 7, E IVP39Dem.]
2. An individual body's nature is its activity [by E IIIPost.1]
3. An individual body's activity produces its existence [by E IIIP6]
4. Therefore, an individual body's ratio produces its existence [by 1, 2, and 3, substituting equivalents]

A more direct, although less informative, route to the same conclusion is supplied by the proposition following E IIIP6. Spinoza argues at E IIIP7 that a thing's striving to exist is its actual essence, a point that is very nearly implied by the definition of 'essence' at E IID2 but which, in its use of the qualification 'actual' and its invocation of perseverance, more closely relates to those things that do exist.[13] Essence, as we have seen, associates closely with nature and form in the *Ethics*. This association is most evident, perhaps, in the extract from the Preface to *Ethics* IV that I quoted above. Perhaps this sentence from the demonstration to E IIP24 is more direct, in that here Spinoza calls essence the ratio that elsewhere he calls nature or form: 'the parts composing the human Body pertain to the essence of the Body itself only insofar as they communicate the motions to one another in a certain, fixed manner [*ratio*] (see the definition after L3C)'. This association suggests clearly that striving, for a body, is its ratio.

These passages, together with the first premise, supply another basis for the argument to the conclusion that an individual body's ratio produces its existence.

1. An individual body's ratio is its nature [by E IIL5 & 7, E IVP39Dem.]
5. An individual body's nature is its essence [by the association of nature and essence, e.g., IV Pref., IVP24Dem.]
6. An individual body's essence produces its existence [by IIIP7, specified for body]
7. Therefore, an individual body's ratio produces its existence [by 1, 5, and 6, substituting equivalents]

The introduction of the notion of power of acting for finite things at the beginning of *Ethics* III, while it is critical to maintaining the distinction between finite things and substance, also obscures the argument. Either of these chains of association, however, suggests that corporeal ratio is a thing internal to an individual body in virtue of which it exists. This, however, is just what a reason is, in Spinoza's account of reasons: that in virtue of which a body exists. Spinoza, moreover, applies many of the same notions – power, perfection, and being – to his account of the natures of finite things that he applies to his account of the nature of substance at E IP11. Recall that in his account of sufficient reason there, Spinoza takes the existence of substance to follow from its nature. So, one of his few examples of sufficient reason invokes the nature of God, in its absolute power and absolute perfection as a reason for absolute existence. Corporeal ratios are the natures of finite things. It seems to me that Spinoza's introduction of the notions of power and perfection into his account of the nature of finite things reinforces the claim that he takes those natures to be similarly, although defeasibly and finitely, reasons for their existences.

Conclusion

Spinoza's use of *ratio* to describe the essence of finite bodies in his definition of the individual and elsewhere is not a mere play on words. His strong, universal claim about reasons at E IP11Dem.2 commits him to the views that all existents have a reason and that the reason for bodies is corporeal. So he owes readers an account of what a corporeal reason is. There is, however, only one genuine explanatory force and only one kind of explanatory force in Spinoza's metaphysics, and those are God and God's power as a self-explanatory thing. Individual things, E IIIP6 makes clear, just are finite expressions of God's explanatory force. Any particular thing, that proposition suggests, will be like God, but finite: a finite self-cause. Where we turn to an account of how particular bodies cause themselves, Spinoza's physics suggests that they do so by means of their own peculiar reason.

To return, finally, to E IP11Dem.2, it seems clear (to me, at any rate) in retrospect that finite things have an external reason in the sense that the reason for their existence is not fully contained in their nature. If you or I or a particular crow exist, it is in large part because the order of corporeal nature permits it. As Spinoza puts the point in the demonstration, nothing prevents or impedes (the Latin is *impedit*) God from existing. Nothing could because there is nothing outside of God. God's nature to cause itself therefore must be efficacious. Things can impede us, however, and we are fortunate if and for as long as they do not. That point, however, does not mean that we do not supply a reason for our own existence. We do, and nature does not supply any other kind of reason.

Notes

1. Della Rocca (1996: ch. 8) is the classic statement of this and related positions.
2. I intend the argument of this chapter to appeal to readers of Spinoza who have a variety of views about what Spinoza means by 'reason' (*ratio*). For a complete account of my view, see LeBuffe (2017).
3. I do not defend the identity of reasons and causes here, because it is widely accepted. For a recent defence of the view, see Newlands (2010). For a recent defence of the view that reasons for bodies are themselves corporeal, see Della Rocca (2012).
4. A detailed account of Leibniz's response to Spinoza is Laerke (2008).
5. I use Curley's translation of the *Ethics* (Spinoza 1985).
6. These are themes of E IApp. In the *Ethics*, Spinoza denies altogether that God (E IP32C1) or human minds (E IIP48) have free will.
7. Spinoza's determinism is most fully expressed at E IP28. For a recent interpretation of determinism and the place of finite modes in the order of things, see Shein (2015).
8. Peterman (2014) is a recent, well-informed but also original and controversial introduction to the physical discursus in the *Ethics*.
9. A notable passage supporting the alternative interpretation is the demonstration to E IIIP7. This is an issue that requires further extended discussion. My conviction that the power of a thing can change while its essence does not, on Spinoza's account, is principally based upon his accounts of the human affects, many of which are such changes.
10. This last option, however, Spinoza takes to be impossible for a finite body. At E IVA1, he asserts that there is no singular thing that cannot be destroyed by another, more powerful singular thing.
11. For accounts of the perfection and reality of finite things generally, see the end of E IVPref., including the passage quoted in the main text here as well as the lines following it, G II 209 1–10.
12. A reasonable concern that a reader might have with this conclusion is that Spinoza emphasises perseverance in being rather than coming into being in these passages. I am convinced that this is not a genuine distinction in Spinoza's metaphysics. Just as God does not cause God, temporally, to come into existence at a particular point, so the power of perseverance in being is not distinct from some different power by which a particular thing comes into being. Spinoza is clearest about this issue at TP 2.2: 'the same power that [natural things] need to begin to exist they need to go on existing'. Thanks to Beth Lord for raising this issue.
13. Here is E IID2: 'I say that to the essence of any thing belongs that which, being given, the thing is necessarily posited, and which, being destroyed, the thing is necessarily destroyed; or this without which the thing can neither be nor be conceived, and vice versa.'

3

Ratio and Activity: Spinoza's Biologising of the Mind in an Aristotelian Key

Heidi M. Ravven

If the parts composing an individual become greater or less, but in such a proportion that they all keep the same ratio of motion and rest to each other as before, then the individual will likewise retain its nature, as before, without any change of form. (E IIL5)

If we compare the later account of ratio and proportion[1] in the *Ethics* with the earlier one of the *Short Treatise on God, Man, and his Wellbeing* we find two things. First, the *Ethics* includes within thinking features of proportion that had been attributed to the body alone in the early work. Second, the *Ethics* account characterises the infinity of Nature (and not just the particulars within it) in terms of 'proportion', namely, as an ordered and nested infinite system composed of subsystems. Spinoza has introduced the homeodynamic force of proportion that maintains the individuality – identity and persistence – of bodies in the *Short Treatise* into his accounts of the operations of the mind in the *Ethics*. And he has rethought embodied infinite Nature in the terms that he had formerly reserved for the coherence of singular bodies.

The maintenance in each of us and in all things great and small of a dynamic and unique individual equilibrium of motion and rest is captured by Spinoza in the notion of maintaining a ratio, a thing's proportion. In the early *Short Treatise* Spinoza had attributed this dynamic self-preservation to the body alone, mind merely passively figuring the dynamic of the body. In the later *Ethics*, however, mind has come to be understood in the proto-biological terms formerly reserved for body alone. For here, the self-maintenance and self-ordering energy of the individual body, its ratio, has come to include and be identified as well with a life-sustaining desire for mental self-perpetuation and self-furtherance – the *conatus* of the mind. In the *Ethics* Spinoza offers an account of thinking as a dynamic, ongoing, and integrative mental *activity* which has various degrees of well-functioning, so that he comes to biologise the mind more fully as he accounts for its functioning in the homeodynamic terms of ratio that he proposed for the singular body alone in the early dualistic theory of the *Short Treatise*. Ratio in the *Ethics* has come to be invoked to account for both the individuality and coherence of particular bodies and also for their expansion within nested larger extensional organisations or systems to infinity. And the *conatus* of the mind is now characterised by features formerly attributed only to the force of ratio-proportion in bodies, and is that which accounts for both the individuality and coherence of particular minds and also for their reaching out to integrate the infinity of their constitutive causes. Spinoza has come to describe the human person as essentially a psychophysical homeodynamic equilibrium that reaches beyond the bounds of the individual, towards internal integration of – and finally coming to be nested within – larger individuals, and ultimately within nature as a whole.

In the *Ethics* Spinoza comes to employ features of bodily ratio-proportion to reconceive the mind in the proto-biological terms that he had previously reserved for body alone. At the same time, bodily ratio is reconceived in the *Ethics* so that it is expressive of infinity and eternity in the account of nested individuals to infinity – an account missing from the *Short Treatise*. Ratio as the internal power of the self-ordering 'proportion' or internal coherence of bodies, and also of the nesting of bodies within ever larger bodies and ultimately within universal nature, has come to have a psychic – and psychological – expression and counterpart as well. For it designates an equally active and biological process of the homeodynamic striving of mind towards self-maintenance, coherence, and also its push towards universality. While the body integrates changing environmental conditions into itself, and integrates itself within (and as) its infinite, nested natural expanse in its very striving to maintain its own proportion or ratio, when *functioning optimally* the mind also actively *integrates* into its self-ordering dynamic process its expanding and unified account of its own causes – causes that ultimately encompass, as internal source and personal constitution, the entire universe. This chapter will trace the development of Spinoza's notion of ratio as proportion from the *Short Treatise* to the *Ethics*.

Spinoza's Early Account of Bodily Proportion in the *Short Treatise*

Proportion, for Spinoza, is the unique *ratio*, the internal relations, that individual bodies maintain and express, which enables them to both constitute and retain their identities amid the constant forces and onslaughts of change by which they are affected. Ratio is the source of their coherence as the singulars they are even when 'other bodies act on ours' (KV IIPref.). That Spinoza is describing, in ratio, a dynamic steady state we learn when he tells us that an 'individual so composed [of others] retains its nature ... so long as each part retains its own *motion*, and communicates it, as before, to the others' (E IIL7). In the *Ethics* ratio also enables these singular bodies to form and become composite individuals, which in turn are nested within larger such individuals, finally coming to be components within the infinity of the one individual, Nature, which is also governed by the force of ratio-proportion. Because of the operation of proportion, the force of maintaining steady ratio (which in contemporary terms we could call 'dynamic equilibrium'), the whole of nature, Spinoza famously says, can be understood to be one such individual. Spinoza writes in the *Ethics*:

But if we should now conceive of another, composed of a number of individuals of a different nature, we shall find that it can be affected in a great many other ways, and still preserve its nature. For since each part of it is composed of a number of bodies, each part will therefore (by L7) be able, without any change in its nature . . . to communicate its motion more quickly or more slowly to the others.

But if we should further conceive a third kind of individual, composed [NS: of many individuals] of this second kind, we shall find that it can be affected in many other ways, without any change of its form. And if we proceed in this way to infinity, we shall easily conceive that the whole of nature is one individual, whose parts, i.e., all bodies, vary in infinite ways, without any change of the whole individual. (E IIL7S)

In the earlier *Short Treatise on God, Man, and his Wellbeing*, however, Spinoza does not put forth an account of composite bodies nested in nature to infinity as he does in the *Ethics* – and this despite his reference in passing in the *Short Treatise* to 'a striving we find both in the whole of Nature and in particular things' (KV I.5). Instead, ratio-proportion is explanatory only of particular bodies, for 'each thing in itself has a striving to preserve itself in its state, and bring itself to a better one'. 'Each and every particular thing that comes to exist becomes such through motion and rest . . . The differences between [one body and another] arise only from the different proportions of motion and rest' (KV II.2Pref.). In the early work, Spinoza also refers to body conceived in Cartesian terms as dominated by 'animal spirits', whose balanced proportion of motion and rest can be disrupted by external forces when love of body dominates the soul.

The soul . . . has the power to move the [animal] spirits where it will; but this power can nevertheless be taken from it, as when, through other causes, arising from the body in general, the proportion [of motion and rest] established in the spirits is taken from them or changed; and when the soul becomes aware of this, a sadness arises in it . . . This sadness results from the love and union it has with the body . . .

[T]he sadness can be relieved in two ways: either by restoration of the spirits to their original form . . . or by being convinced by good reasons to make nothing of this body. The first is temporary . . . But the second is eternal, constant, and immutable. (KV II.20)

In the *Short Treatise* Spinoza does not yet regard the process of mental life from the biological perspective that he comes to in the *Ethics*, or in the way that he regards bodily coherence in both texts. In the earlier treatise, Spinoza tellingly refers quite often to the soul's fundamental conflict between longing for mental transcendence of the body and its competing love of its body:

[I]f other bodies act on ours with such force that the proportion of motion [to rest] cannot remain 1 to 3, that is death, and a destruction of the soul, insofar as it is only an Idea, knowledge, etc. of a body having this proportion of motion and rest.

However, because it is a mode in the thinking substance, it has been able to know and love this [substance] also, as well as that of extension; and uniting itself with these substances (which always remain the same), it has been able to make itself eternal. (KV IIPref.)

Hence in the *Short Treatise* the human soul is fraught with internal conflict that can be resolved only by favouring one love and union over the other, for the choice of one represents the distancing and disunion from the other. This is in contrast to the single striving for self-preservation and self-furthering that is expressed in and characterises mind and body in the later *Ethics*. Spinoza's early thought portrays the soul as poised between two great loves and two potential unions, either with one's own body or with the divine.

Those steeped in the *Ethics* might find the dualisms and traditional otherworldliness of Spinoza's early thinking in the *Short Treatise* shocking. For he writes there of a human person whose very nature is drawn to and poised between conflicting and mutually exclusive desires and realms.

Soul can be united either with the body of which it is the Idea, or with God, without whom it can neither exist nor be understood. From this, then, one can easily see that:
1. if it is united with the body only, and the body perishes, then it must also perish; . . .
2. if it is united with another thing, which is, and remains, immutable, then, on the contrary, it will have to remain immutable also. (KV II.23)

Accordingly, the soul, already defined as the Idea of the Body as it is in the *Ethics*, nevertheless is faced with a choice between union with the ephemeral, natural, singular body narrowly conceived or union with the non-corporeal, infinite, and eternal God. The human person is suspended and torn between opposing realms – a dualism that Spinoza will, of course, resolve in the *Ethics*, in part through the further development and application of the ratio-proportion theory. The depth of Spinoza's commitment to ontological and intra-psychic dualism in the *Short Treatise*, however, is striking:

We shall now try to explain the union we have with [God] by Nature and by love. We have already said that there can be nothing in Nature of which there is not an idea in the soul of the same thing. And as the thing is more or less perfect, so also are the union of the Idea with the thing, or [instead] with God himself . . .

And because the body is the very first thing our soul becomes aware of – for as we have said, there can be nothing in Nature whose Idea does not exist in the thinking thing, the idea which is the soul of that thing – that thing must, then, necessarily be the first cause of the idea.

But this idea cannot find any rest in the knowledge of the body, without passing over into knowledge of that without which neither the body nor the Idea itself can either exist or be understood. Hence, as soon as it knows that being, it will be united with it by love.

To grasp this union better and infer what it must be, we must consider the effect [of the union] with the body. In this we see how by knowledge of and passions toward corporeal things, there come to arise in us all those effects which we are constantly aware of in our body . . . and so (if once our knowledge and love come to fall on that without which we can neither exist nor be understood, and *which is not at all corporeal*) the effects arising from this union will, and must be, incomparably greater and more magnificent. For these effects must necessarily be commensurate with the thing with which it is united.

When we become aware of these effects, we can truly say that we have been born again. For our first birth was when we were united with the body . . . But our other, or second, birth will occur when we become aware in ourselves of the completely different effects of love produced by knowledge of this incorporeal object. *This [love of God] is as different from [the love of the body] as the incorporeal is from the corporeal.* (KV II.22, my emphasis)

It is within this early dualist context that Spinoza introduces the bodily proportion and ratio account of particular bodies.

The Aristotelian and Galenic Background of Spinoza's Account of Ratio-Proportion

Jacob Adler has argued that Spinoza's notion of proportion, understood as ratio, especially in the *Short Treatise*, reflects the influence of both the Aristotelian tradition, particularly the early third-century philosopher Alexander of Aphrodisias, and of seventeenth-century academic medicine and personality psychology, which were still largely within the tradition of Galen. Alexander of Aphrodisias, the most important commentator on Aristotle until the thirteenth-century Averroes, was famous for his radical psychological theory that the soul emerged from the harmony or attunement of bodily parts and functions. Adler shows convincingly that Spinoza could and probably would have had knowledge of Alexander's theory from a number of sources, some of which were in his library. I won't repeat his argument here. Adler goes on to quote a revealing passage in which Alexander writes:

The soul is not a particular kind of blend [*temperamento*] of bodies – which is what harmony is – but a power that emerges above a particular kind of blend [*temperamento*] of bodies analogous to the powers of medicinal drugs, which are assembled from a blend [*immixtione*] of many [ingredients] . . . [T]he mixture, composition and proportion of such drugs – . . . it might turn out is 2:1, another 1:2, and another 3:2 – bear some analogy to a harmony. The power, however, which emerges from the blend of drugs exhibiting the harmony and proportion is not likewise a harmony, too ... The soul is also of this sort. *For the soul is the power and form that supervenes on the blend [temperamento] of bodies in a particular proportion, not the proportion or composition of the blend [temperamento]*. (Adler 2014: 25, my emphasis)

Hence for Alexander, and for Spinoza in the *Short Treatise*, proportion cannot be attributed to the soul itself – a position on which Spinoza comes to change his mind in the *Ethics*. The account in the *Short Treatise*, however, bears a similarity to this passage of Alexander, as Adler shows by reference to the following passage in the *Short Treatise* (KV IIPref.):

6. This knowledge, Idea of each particular thing that comes to exist, is, we say the soul of this particular thing.
7. Each and every particular thing that comes to exist becomes such through motion and rest ...
8. From this proportion of motion and rest, then, there comes to exist also this body of ours, of which ... there must exist a knowledge, Idea, etc. in the thinking thing. This Idea, knowledge, etc., is also our soul ...
12. So if such a body has and preserves its proportion – say of 1 to 3 – the soul and body will be like ours now are; they will, of course, be constantly subject to change, but not to such great change that it goes beyond the limits of 1 to 3; and as much as it changes, the soul also changes each time ...
14. But if other bodies act on ours with such force that the proportion of motion [to rest] cannot remain 1 to 3, that is death, and a destruction of the soul, insofar as it is only an Idea, knowledge, etc. of a body having the proportion of motion and rest.

Adler (2014: 25) concludes that in both Spinoza's *Short Treatise* and Alexander's account, 'the soul comes to be when the elements composing the body are mixed in a certain quantitative proportion'. And of course, the scandalous corollary of the Alexandrian position, not lost on later Jewish philosophers and others in that tradition, was that, on that basis, it could be argued that an individual's soul perishes with the body – a consequence that Spinoza embraces in the *Short Treatise*.

The application of ratio-proportion to psychic (as well as physical) functioning was in the air at the time as well, according to Adler. Galenic medicine was based on harmonising the four humours (blood, phlegm, black bile, and yellow bile), which were also the bases of personality types (sanguine, phlegmatic, choleric, and melancholic). So the four humours provided not only an account of illness in terms of bodily imbalances but also a character psychology. Both health and mental health depended upon harmonising psychophysical components. Adler points out that we now know that Spinoza attended lectures in medicine at Leiden. He goes on to recount how Spinoza, in a Cartesian modernising modification, came to reduce the four humours to two, identifying them as motion and rest. Adler then provides evidence and argument that the 'ratio' Spinoza refers to (inspired by Alexander of Aphrodisias) is integrated into a Galenic account of the harmony or balance at the basis of temperament per se and as specified in the particular balances that produce unique temperaments. Yet in the *Short Treatise* there is a feature of the mind, which Adler also identifies as having been adopted from Alexander, that appears to prevent Spinoza from developing a full psychological theory at this stage: it is the *passivity* of the mind in all thinking, whether in the case of its knowledge-love of its body or in that of its knowledge-love of the divine (Adler 2014: 25). '[A] finite intellect can understand nothing through itself unless it is determined by something external' (KV I.1). And later on:

> The intellect is wholly passive, i.e., a perception in the soul of the essence and existence of things. So it is never we who affirm or deny something of the thing; it is the thing itself that affirms or denies something of itself in us. (KV II.16)

Moreover, the passivity of the mind is part and parcel of the broader knowledge-union-love account, so that the soul can achieve immortality only by conjunction or union with the intellect of God – a position held by many in the Jewish philosophical tradition, including Maimonides. In Alexander's version, the human mind is either passive to the body or passive to God. The resulting account of the emotions in the *Short Treatise* suffers from the limitations imposed by this model despite the reference in this context to motion and rest (and by implication thereby to proportion):

> [I]t is easy to infer what are the principal causes of the passions. For regarding the body, and its effects, Motion and Rest, they cannot act on the soul otherwise than to make themselves known to it as objects. And according to the appearances they present to it, whether good or bad, so the soul is also affected by them . . .
> [I]f something else should present itself to the soul more magnificently than the body does, it is certain that the body would then have no power to produce such effects as it now does. (KV II.19)

Hence it is the knowledge-union of the soul with the body, or alternatively with God, that forms the basis of Spinoza's psychological theory of the passions in the early text. Spinoza emphasises that his is not a reductively bodily theory, for the passions do not arise 'insofar as the body is a body', so that 'the body is not the principal cause of the passions'. Instead, the passions arise due to 'the kind of knowledge the soul has each time' (KV II.19). Perhaps Spinoza is implicitly suggesting a critique of – rather than adopting – the then contemporary, Galenic, reductively bodily account of temperament that attributed emotions to dominant chemical mixtures[2] and its Cartesian version based on the 'animal spirits'.[3] It seems that in his zeal to avoid a reductively corporeal account of the emotions, Spinoza has embraced what can only be counted as a mythic theological one, one that was a commonplace of a philosophical tradition in which he was steeped.[4] It is the overly cognitivist ('Knowledge is the proximate cause of all the Passions of the soul'; KV II.2) and theological account of the passions of the *Short Treatise* with which Spinoza will break in the *Ethics*, in the event recruiting and employing and integrating the ratio-proportion theory in a new way, and thereby inventing what he himself regarded as, and argued was, the first truly scientific psychology (E IIIPref.).

Spinoza's Turn to a Quasi-Aristotelian Resolution

In the *Ethics* Spinoza no longer regards the mind, as he did in the *Short Treatise*, as passive in all its cognitive operations, but instead embraces the notion that 'activity' characterises the optimal cognitive state – and bodily state as well. He proposes that '[t]he Mind, as far as it can, strives to imagine those things that increase or aid the Body's power of acting' (E IIIP12). Here Spinoza maintains that the adequacy of a person's ideas and the 'activity' (in contrast with the 'passivity') of one's thinking are co-defined (E IIIP1). Hence, 'it follows that the Mind is more liable to passions the more it has inadequate ideas, and conversely, is more active the more it has adequate ideas' (E IIIP1C). Knowledge is empowering because it is, and is within, one's own power to arrange ideas rationally as explanatory causal series, and hence its order is self-generated, in contrast with the passive acceptance of external imaginative associations that masquerade as explanatory (E IIID1 & 2). And so too is the body capable of activity and passivity:

> The human body can be affected in many ways in which its power of acting is increased or diminished, also in others which render its power of acting neither greater nor less. (E IIIPost.1)

Spinoza further holds that we are 'active' insofar as we are the adequate cause of our actions, that is, when our actions can be understood to follow from our nature alone; and conversely, we are passive when we are their partial cause (E IIID2). Moreover, body and mind express the same state of either relative activity or passivity:

> The idea of any thing that increases or diminishes, aids or restrains, our body's power of acting, increases or diminishes, aids or restrains, our mind's power of thinking. (E IIIP11)

As Susan James points out (1997: 155), 'Because mind and body are the same thing described under different attributes, they can only act or be acted on together.' Spinoza's notion of activity combines the Aristotelian notion of perfection as biological optimal functioning and the theory of ratio as proportion that maintains a dynamic individuality within nested, ever-larger bodies to infinity. He utilises an Aristotelian notion of the mind as active in this biological and naturalist sense of optimally functional while introducing into this model features of the account of body as maintaining internal stability and identity as proportion. Susan James, in *Passion and Action*, traces how philosophers in that era adopted the Aristotelian dichotomy of activity versus passivity while at the same time adapting the model of an active formal principle and a passive matter or material principle to the new mechanistic science. Breaking with the Aristotelian metaphysics and physics of formal and final causes, they nevertheless retained much of the active/passive conceptual framework, especially in theories of the emotions.

Aristotle held that achieving an optimal dynamic state of human well-functioning, which he designated as 'activity', was both the essential characteristic of theoretical thinking and the quintessential mark and aim of human nature and of humanity as a natural species, the rational animal. In the *Nicomachean Ethics*, Aristotle developed a biological model according to which each species, thing, and practice had a natural activity, proper to itself, that defined it and whose maximal functioning defined what it was to be a perfect exemplar of the kind of thing it was. It was a dynamic steady state without any goal outside its own well-functioning at which it aimed, so that means and ends coincided. Aristotle held that all natural things ought to be described in this way (NE X, 5, 1176a3–10 [Aristotle 1970: 1101]). While all natural kinds had defining activities, for human beings that activity was thinking, and in particular theoretical thought rather than practical deliberation.

> [J]ust as for a flute-player, a sculptor, or an artist, and, in general, for all things that have a function or activity, the good and the 'well' is thought to reside in the function, so would it be for man, if he has a function . . . Life seems to be common even to plants, but we are seeking what is peculiar to man . . . There remains, then, an active life of the element that has a rational principle . . . And, as 'life' of the rational element has two meanings, we must state that life in the sense of activity is what we mean . . . human good turns out to be activity of soul in accordance with virtue, and if there are more than one virtue, in accordance with the best and most complete. (NE I, 7, 1097b26–1098a17 [Aristotle 1970: 942–3])

Aristotle goes on to characterise every species as having its own good which is its embedded end and characteristic activity. That is to say, all natural things belong to categories defined by activities of this kind, in which the end is intrinsic to the means (NE X, 5, 1176a3–10 [Aristotle 1970: 1101]). Human beings are defined by such a process or activity. Both the final process and its coming to full actuality define, for Aristotle, the good. The good is, therefore, relative to the end or species. For the human being it is reason.

Spinoza recruits Aristotle's account of well-functioning, which he too terms 'activity' in contrast with 'passivity', but in the *Ethics* eschews the identification of that function with reason alone. Instead, he uses it to explain a psychophysical state of *desire* spanning body and mind and expressed in the various emotions, which are evaluated in terms of the relative dynamism and self-origination of the process that produces them.⁵ For all emotions, passive as well as active, are characterised by their intense drive for self-preservation and promotion – the bodily force of homeodynamic proportion now taken up into the mind's operation of thinking itself.

> [S]ince (by IIP11 and P13) the first thing that constitutes the essence of the mind is the idea of an actually existing body, the first and principal [tendency] of the striving of our mind (by P7) is to affirm the existence of our body. (E IIIP10Dem.)

In the *Ethics*, this desire has come to be the 'very essence' of human beings (E IIIDef.Aff.I), having both mental and bodily expressions, and in both is a striving for self-furthering, and self-protectiveness, and the self-coherence that Spinoza describes in his account of the maintenance of bodily proportion in both the *Short Treatise* and the *Ethics*. Mind as well as body are fundamentally driven by and expressive of the desire for self that is the essence of human beings, and which in the mental arena entails the urge to the preservation of one's own beliefs, attitudes, and ideas in the way that it entails in the physical arena the self-preservation of one's unique 'ratio'.

It is emotion that spans body and mind, according to Spinoza's definition, and hence provides the opportunity for a psychophysical understanding of the human person as a homeodynamic system seeking its own internal coherence, perpetuation, furtherance, and stability. For Spinoza holds that emotion is composed of the bodily affections plus the ideas of those affections (E IIID3). Emotions consist of the three primary affections (pleasure, pain, and desire) (E IIIP11S) plus the ideas of their causes, which causes can be imagined to be external or understood as internal to the order of the mind itself and hence rationally self-generated, that is to say, active. Hence, emotions express the functional state of the person as a whole, each emotion registering the relative condition of the *conatus* of a person as his or her agency, or 'activity', is strengthened or weakened (E IIIP11). The enhancement of activity arising from causal self-understanding is the product of rational self-reflection about the causes of one's own emotions, while the weakening of the *conatus* stems from reactive submission ('passivity') to the imaginative associations dictated by external circumstances, attitudes, and beliefs.

Rendering the emotions active entails the progressive *inclusion* and rational ordering of all one's causes (to infinity) within the self-ordering, ratio-maintaining activity of the *conatus* of the mind – an account and elaboration that Spinoza, not Aristotle, brings to the table – while passivity entails the potential of unintegrated external mental associations and memories to dominate – render passive – the mind's thinking and feeling state and relative condition. The workings of the force of proportion in bodies was described in just these terms as operating to maintain internal self-order of its own ratio-coherence amid the onslaught of external forces and in the face of the danger of passively succumbing to them and hence disintegrating. We recall KV IIPref.:

if such a body has and preserves its proportion – say of 1 to 3 – the soul and body will be like ours now are; they will, of course, be constantly subject to change, but not to such great change that it goes beyond the limits of 1 to 3 . . . But if other bodies act on ours with such force that the proportion of motion [to rest] cannot remain 1 to 3, that is death.

Hence Spinoza's account of bodily proportion – that is, of the internal self-ordering and self-stabilising energy or force that maintains its own individuating identity and coherence amid the external onslaughts of life, by integrating them and being integrated into them both physically and mentally – can help us understand his account of emotion. Spinoza's rethinking of mind as *conatus*, with the capacity for achieving an active and integrative psychophysical engagement with the world, captures and expresses mentally the dynamic character of the homeodynamic self-preservation of objects, which the earlier account of emotion as characterised by two forms of cognitive passivity and the choice between two loves of the *Short Treatise* did not. Spinoza has not yet in the latter, it seems, come up with an account that fully biologises the mind, while nevertheless maintaining its mental character. In the earlier work, the mind is passive to the body, suggesting a kind of proto-psychobiology reminiscent of Descartes' *Passions of the Soul*; but the mind is also capable of great independence from the body in its figuring of universal content as the love and knowledge of God. The *Ethics*, by contrast, captures the person as a single process, a single central desire for self-preservation and self-furthering described in two ways, physically and mentally, both expressing the common character of the dynamic self-maintenance and self-ordering coherence of the *conatus* as both mental and physical. Hence in his theory of the mind and body *conatus*, Spinoza combines aspects of Aristotle's biological theory of wellbeing, as the maximal mental functioning, with the homeodynamic force for stability and maintaining identity of the bodily proportion theory. He thereby replaces the earlier mythic account of the two opposing loves with a thoroughly naturalistic and proto-biological – but not reductionist – theory of mind and emotions.

Furthermore, the expansion in the *Ethics* of the description of ratio-proportion to include the nesting of component individuals in ever-more encompassing individuals indicates that Spinoza has also transformed the meaning of ratio to explicate the causal structure of nature itself, and hence what the mind comes to know and embrace in tracing its own causal constitution to and within infinity. Ratio as the force of proportion in body and ratio as reason – the dynamic, active process of the mind – have become one insofar as the active tracing of causes of the self, which is the mental operation of adequate and active thinking, traces one's own causal constitution within those nested organisations and dynamic self-organising systems. Yet the theory of ratio as developed in the *Ethics* as active mental *conatus* also provides a way to account for and maintain the mental *individuation* of subjects (of minds), and not just the physical individuation of objects. Hence Spinoza has overcome a major problem of the *Short Treatise*, namely that mental union with the divine there implies the subsuming of mental individuation in universality and an ultimate overriding of the individual mind. In the *Ethics*, the individuation and coherence of each mind can be maintained even as it expands its content to infinity because Spinoza has introduced the operation of maintaining individuating proportion within the mental operations of the singular mind.

A Final Word on Ratio as Pointing towards Systems Theory in the *Ethics*

Spinoza's nature is a biological system in that it is an ordered totality composed of a hierarchy of subsystems, each more complete and independent than its components. I have argued elsewhere that Spinoza has envisioned a systems theory approach to explaining the universe and the human person *avant la lettre*. A systems theory may be understood in terms defined by Berrien:

> A system is defined as a set of components interacting with each other and a boundary which possesses the property of filtering both the kind and rate of flow of inputs and outputs to and from the system. It has been customary to distinguish between open and closed systems: open systems are those which accept and respond to inputs (stimuli, energy, information and so on) and closed systems are those which are assumed to function within themselves. (Berrien 1968: 3)[6]

Spinoza's notion of activity, insofar as it combines the Aristotelian notion of perfection as biological optimal functioning and the theory of ratio as proportion that maintains a dynamic individuality within nested, ever-larger bodies to infinity, presages the theory of the self-ordering and self-integrating process of an open system encompassing everything. *Hence all of nature is self-ordering energy or a process of ordering and integration.* All of nature is the process of its self-becoming, or, more precisely, of its active and necessary self-being (E IP29). Spinoza has developed a model of an infinite system whose goal and activity is perfected self-generation, self-perpetuation in dynamic equilibrium or a steady state. Psychology and ethics are thereby reinterpreted by Spinoza in terms of the bi-directionality of each of us as a nested subsystem within the natural infinite system. The first direction is the integration into one's mental *conatus* of all one's constitutive causes to infinity – internal integrity. The second direction is externalising or extruding that integrated coherent self into the ongoing self-ordering proportion that maintains the coherence of the infinity of nature. For the individual *conatus* has achieved identification with the natural universe, insofar as it is now inside oneself as one's own explanatory causes. Personal wholeness is accomplished only in universal identification and vice versa.

Notes

1. In this chapter I use ratio to refer to proportion, to a thing's unique equilibrium, unless explicitly designated otherwise.
2. My proposal here that Spinoza is rejecting and implicitly criticising the Galenic temperament theory of psychology challenges Adler's claim (2014: 29) that Spinoza is adopting it in this passage and chapter. We both agree, however, that he is making reference to it.
3. Curley (Spinoza 1985: 99) suggests that Spinoza is critiquing the Cartesian account here, in his note 2 to KV II.2.
4. Rather than this account being particular to Alexander of Aphrodisias alone, Alfred Ivry (2009: 54) has argued that it is a commonplace of the entire philosophical tradition in question. Ivry calls it 'the dominant epistemological paradigm of the period as laid out by al-Farabi, Avicenna before him, and by Alexander of Aphrodisias in the third century'. And Maimonides subscribed to it as well, a source that Spinoza was very familiar with.
5. See Ravven (1990) for an extended treatment of the primacy of desire over cognition in Spinoza's philosophy.
6. See Berrien (1968: 14, 15, 17) where he defines a theory of system that is to cover biological, physical, chemical, mathematical, psychological, social, and other systems. See also Ravven (1989) and my several treatments of Spinoza's philosophy as systems theory (Ravven 2004; 2005; 2012; 2013), and Zimmermann (2010).

4
Harmony in Spinoza and his Critics

Timothy Yenter

Men have been so mad as to believe that God is pleased by harmony. Indeed there are philosophers who have persuaded themselves that the motions of the heavens produce a harmony. All of these things show sufficiently that each one has judged things according to the disposition of his brain; or rather, has accepted affections of the imagination as things. (E IApp.)

Things which are of assistance to the common society of men, or which bring it about that men live harmoniously, are useful; those, on the other hand, are evil which bring discord to the state. (E IVP40)

Spinoza is in a potentially untenable position. On the one hand, he argues that those who claim to see harmony in the universe are badly mistaken; they are imagining falsely rather than reasoning properly. On the other hand, harmony is positively discussed in his ethical writings and even serves as the basis for his vision of society. How can both be maintained? In this chapter I argue that this *prima facie* conflict between the two treatments of harmony is resolvable, but that in resolving it a new set of questions for Spinoza is raised. The focus is on the *Ethics*, but the issues carry into the *Political Treatise* and various letters as well, so I will address them as they become relevant.

The chapter proceeds in three stages. In the first part, I introduce the *prima facie* conflict for Spinoza. Although he roundly criticises harmony, most notably in the Appendix to the first part of the *Ethics*, he also endorses a particular understanding of harmony that he relies on in his metaphysical, ethical, and political writing. I argue that there are two distinct concepts at work and that the *prima facie* conflict dissolves once this is recognised. In the second part, I show that the second (positive) concept in Spinoza generates a new set of questions. To accomplish this I explore how *harmony* was used in the sixteenth to the eighteenth centuries, which will help us better situate Spinoza. In doing so, we can recognise new possibilities and new connections between Spinoza and philosophers who are not often discussed in connection with him. In the final section, I explore some of these new connections to reveal how they connect to recognised problems of Spinoza interpretation.

The Internal Conflict

Seventeenth- and eighteenth-century writers frequently used harmony as a theoretical and rhetorical tool. For instance, the British Rosicrucian philosopher John Heydon writes:

> Moreover also from the proportions of the Motions of the Planets amongst themselves, and with the eighth Sphere resulteth the sweetest Harmony of all ... But by the violent motion of the *Primum Mobile* is the most swift and accute [sic] sound of all; but the violent Motion of the *Moon* is most slow and heavy, which proportion and reciprocation of motions yeilds [sic] a most pleasant Harmony; from hence there are not any Songs, Sounds, or Musicall [sic] Instruments more powerful in moving mans [sic] affections or introducing impressions, then those which are composed of Numbers, Measures and Proportions, after the example of the *Heavens*; Also the *Harmony* of the *Elements* drawn forth from their basis and *Angles*, I shall speak of in order ... (Heydon 1662: 51–2)

Heydon claims that there is a harmony in nature that derives from the mathematical proportions between the motions of the planets, that this harmony is pleasing to us, and that there can be further harmonies between the elements, suggesting that harmonies exist throughout nature in various ways. His view is similar to what the alchemists say about harmony and what Johannes Kepler and other astronomers say about the universe.

Jump forward forty-three years to the Newtonian philosopher and theologian Samuel Clarke, who says, in the first set of his Boyle lectures,

> Yet the Notices that God has been pleased to give us of himself, are so many and so obvious; in the Constitution, Order, Beauty, and Harmony of the several Parts of the World; in the Frame and Structure of our own Bodies, and the wonderful Powers and Faculties of our Souls; in the unavoidable Apprehensions of our own Minds, and the common Consent of all other Men; in every thing within us, and every thing without us: that no Man of the meanest Capacity and greatest Disadvantages whatsoever, with the slightest and most superficial Observation of the Works of God, and the lowest and most obvious attendance to the Reason of Things, can be ignorant of *Him*; but he must be utterly without Excuse. (Clarke 1738: 2.576–7)

Clarke, like Heydon, posits a harmony to the universe, but Clarke parlays this claim into an argument that the harmony and order of the universe must be explained by the work of a transcendent divine being. Through reason's operation on the information we take in about the universe, we recognise that a divine being freely created the universe.

I know of no reason to think that Spinoza read Heydon, and he certainly didn't read Clarke, who lived later, but he was keenly aware of both of these families of claims. His response to both groups is rooted in identifying what he believes to be their common mistake. He lays this out most clearly in the appendix to Part I of the *Ethics*.

> And since those things we can easily imagine are especially pleasing to us, men prefer order to confusion, as if order were anything in nature more than a relation to our imagination. They also say that God has created all things in order, and so, unknowingly attribute imagination to God – unless, perhaps, they mean that God, to provide for human imagination, has disposed all things so that men can very easily imagine them . . .
>
> [T]hey believe all things have been made for their sake, and call the nature of a thing good or evil, sound or rotten and corrupt, as they are affected by it . . . Men have been so mad as to believe that God is pleased by harmony. Indeed there are philosophers who have persuaded themselves that the motions of the heavens produce a harmony. All of these things show sufficiently that each one has judged things according to the disposition of his brain; or rather, has accepted affections of the imagination as things . . .
>
> We see, therefore, that all the notions by which ordinary people are accustomed to explain nature are only modes of imagining, and do not indicate the nature of anything, only the constitution of the imagination. (E IApp.)

Spinoza's response seems clear enough. Harmony is often posited as a feature of the world, when it is actually just an affection of the imagination. Not only do humans not all find the same things harmonious, but some even foolishly say that God finds those things harmonious, falsely attributing imagination to God and arrogantly presuming that God created things for our enjoyment. Harmony is based on the particular bodily constitution of the thing affected pleasingly, and these constitutions differ, so we ought not claim that perceived order or harmony is a deep feature of the world or provides us any insight into the nature of God.

With this response to the harmonisers in mind, my purpose is to raise two problems for Spinoza's response. The first problem, which I call 'the *prima facie* conflict', is that although Spinoza speaks dismissively of harmony in the appendix to Part I of the *Ethics*, harmony plays a fundamental constructive role in his civil ethics in Part IV. How can Spinoza chastise those who attribute perceived harmonies to the universe and to God but also appeal to harmony in his account of a just community? I argue that the positive notion is a recognition of equivalence grounded in a metaphysical relation of similarity or identity of nature, while the other incorrectly presumes that perceived harmony extends to beings other than us on no good metaphysical basis. This resolution to the conflict can succeed once we realise that Spinoza equates true harmony (*concordia*) with identity or similarity and false harmony (*harmonia*) with pleasingness. The second problem I consider is whether Spinoza is justified in maintaining this simple distinction between harmony as similarity or identity and harmony as a real feature of the universe in light of the many different ways that harmony is used by his near contemporaries.

Harmony was a theoretically powerful and frequently discussed concept for thinkers of many traditions and methods. Harmony, though, is not a single concept, but a cluster of related concepts used by astronomers describing the movement of celestial bodies, by alchemists exploring the sympathies between microcosms and the macrocosm, by aestheticians explaining the qualities that give rise to beauty, by anatomists describing the workings of the body, and by metaphysicians proposing ersatz unities of composites.[1] Later, I will detail some of the concepts that are labelled harmony, but for now we can begin with an apparent conflict between Spinoza's two approaches to harmony.

Spinoza uses harmony in at least two ways. Living harmoniously with others is a key theme of Spinoza's ethical and political writing (for example, E IVP35S2, E IVP40, E IVApp.). For instance, 'Only insofar as men live according to the guidance of reason, must they agree in nature' (E IVP35). Here, Spinoza discusses *concordia* (harmony) in connection to justice and living in community. However, ignorant people think that what is pleasing to them, such as *harmonia* (harmony), is also pleasing to God (E IApp.). The *prima facie* conflict is that Spinoza seems both to criticise and to positively employ harmony in his writing. This is not limited to the *Ethics*, but it does find a particularly clear formulation there. We could also look outside the *Ethics*, where we find that *concordia* is used in a positive sense in the *Political Treatise* (TP 6.4).

Can the *prima facie* conflict be resolved? I believe it can, although we will discover new problems as we do so. Let me first stress that we should not resolve the conflict simply by noting that Spinoza uses the Latin *concordia* when talking about harmony in a positive sense and the Latin *harmonia* when using harmony in a negative sense. The use of two different terms may appear to suggest that Spinoza recognises some difference that grounds two different concepts, but it does not automatically do this. (Compare: it is not enough simply to use the words 'free' and 'voluntary' differently without some sort of grounding or theoretical account to distinguish the concepts.)

Spinoza has at least two conceptions of harmony in the *Ethics* and elsewhere. One thing that marks the difference in uses is the conceptual difference between *pleasingness* and *similarity*. Harmony in the negative sense is what a person finds pleasing; harmony in the positive sense is what has the same or similar nature to me and therefore is useful to me. What is pleasing can be useful and what is useful can be pleasing, so these are not exclusive categories, so we have not yet shown that the two are in fact distinct concepts. Spinoza needs to offer some further grounding – something like *merely pleasing* versus *actually similar* – and explain why this difference matters.

Let's look again at Spinoza's complaint in the appendix to Part I of the *Ethics*. Spinoza chastises people who believe that the apparent order of the universe reflects the actual state of the universe, when it merely reveals their own imagination. They attribute imagination to God when they say that God created this order (or alternately, that God established something that would seem orderly to us, thus giving us a false pride of place in the universe). This is extended to discussions of good and bad, beautiful and ugly. He then turns to harmony:

[T]hose which move the ears are said to produce noise, sound or harmony. Men have been so mad as to believe that God is pleased by harmony. Indeed there are philosophers who have persuaded themselves that the motions of the heavens produce a harmony. All of these things show sufficiently that each one has judged things according to the disposition of his brain; or rather, has accepted affections of the imagination as things. (E IApp.)

Mistaking 'the affections of the imagination' for actual things is the mistake that the harmonists make. Their initial mistake is to believe that because they find a thing pleasing, the thing really is pleasing. Because they believe the thing to be really pleasing, they conclude that God, too, must find it pleasing. Spinoza denies that pleasing-to-me entails pleasing-in-itself. By blocking this entailment, he blocks the conclusion that what is pleasing-to-me is also pleasing-to-God. Relatedly, to agree (*convenire*) with the intellect is the mark of a true idea in *The Treatise on the Emendation of the Intellect*, while the imagination is acted on according to laws other than the laws of the intellect (TIE 86).

Spinoza claims that this mistake is responsible for many controversies and ultimately for scepticism. To avoid these controversies and the resultant scepticism, we should recognise that 'although human bodies agree in many things, they still differ in very many. And for that reason what seems good to one, seems bad to another; what seems ordered to one, seems confused to another; what seems pleasing to one, seems displeasing to another, and so on' (E IApp.). The differences between human bodies should be enough for us to recognise that what is pleasing-to-me is not necessarily pleasing-to-another. Spinoza makes clear that whether something is pleasing-to-me is grounded in features of my body. This was introduced by his discussion of the senses a few sentences earlier[2] and is reiterated by the appeal to the differences between bodies here. It is bodily difference that explains why I cannot conclude that something is pleasing-to-another on the basis that it is pleasing-to-me; this is magnified when the gap is between a human being (me) and God. The move to God is also illegitimate because it asserts that God has imagination and sensation.

On Spinoza's first conception of harmony, harmony is pleasingness; because this pleasingness is based on sensations in the body, and bodies differ, we cannot conclude that what is harmonious to one is harmonious to another, especially when we move from one sort of body to another. On his second conception, harmony is the agreement of another's nature with one's own and the recognition of this agreement. Spinoza discusses agreement with one's nature in two ways. In Part IV of the *Ethics*, he is concerned with the agreement of a thing with *human* nature. '*Insofar as a thing agrees [convenit] with our nature, it is necessarily good*' (E IVP31). If two human beings are subject to their passions, then they are not in harmony (E IVP32), but if they live according to reason, then they agree in nature (E IVP35) because human nature is rational. Importantly, this harmony is the basis for the ideal civil society.

The agreement with human nature is a special case of a general point that Spinoza makes about harmony with a being's nature. In Ep. 32 to Henry

Oldenburg (the 'worm in the blood' letter), Spinoza explains what he means by the coherence of parts, which is cashed out in terms of opposition and adaptation.

> Concerning whole and parts, I consider things as parts of some whole to the extent that the nature of the one adapts itself to that of the other so that they agree with one another as far as possible. But insofar as they disagree with one another, to that extent each forms in our Mind an idea distinct from the others, and therefore it is considered as a whole not as a part.

Spinoza uses the evocative example of a worm living in an animal's blood. What the host animal would consider parts of itself, the worm would consider to be wholes. The difference is that from the host's perspective, the 'chyle, lymph, etc., are considered as parts of the blood', but the worm in the blood 'when it encounters another, either bounces back, or communicates a part of its motion' so it 'would consider each particle of the blood as a whole, not a part' (Ep. 32). Spinoza's point here is only partly about the relativity of perspective. The host animal's perspective includes 'how all the parts of the blood are regulated by the universal nature of the blood' (Ep. 32). This knowledge is superior to the worm's perspective because it appreciates the 'universal nature' of something. Presumably, an even higher form of knowledge would be to understand how the whole universe is regulated; this is confirmed by *Ethics* IIP40S2, where the highest form of knowledge is claimed to be that which moves from the adequate idea of attributes to adequate knowledge of the essence of things. Humans are capable of such intellectual knowledge, which allows us to know things *sub specie aeternitatis*. Living with this intellectual apprehension of how things properly organise is central to Spinoza's conception of ideal individual life, but since this is every human's rational nature, it is also the ideal of the communal life, because it is our reason and not our bodies that marks us as human. The discussion in the second half of *Ethics* IV is a version of this general point that the less distinct two things are, the more they are in harmony. Given that living in harmony is, for Spinoza, living by reason (TP 5.5), his positive conception of harmony is grounded in the nature of human beings, our reason. We imagine, too, but because affectations of the imagination differ due to our dissimilar bodies, he is dismissive of harmony understood as pleasingness.

With these two conceptions of harmony before us, we can see how Spinoza overcomes the *prima facie* conflict. If by harmony we mean what is pleasing to us, we cannot infer from the fact that something is pleasing to us that it is pleasing to God or that it reveals something important about the universe. Spinoza treats this conception of harmony negatively because it is the basis for these two frequent mistakes. However, harmony understood as the agreement of two rational natures is treated positively because it is grounded in an actual sameness or similarity of two things. While harmonious pleasingness cannot be attributed to others because of bodily *difference*, harmonious agreement is attributable to others insofar as there is *agreement* in their rational natures. This is how Spinoza overcomes the *prima facie* conflict.

What seems at first like a conflict turns out to be two (or more) distinct notions of harmony in Spinoza. Once we note these uses and how they are distinguished, we not only overcome the *prima facie* conflict, but we can recognise that Spinoza has a ready defence for those later critics, like Samuel Clarke, who would run the argument that the beauty and order and harmony of the universe are evidence that God created it freely and not out of necessity. Spinoza is often singled out as the target of these arguments, but distinguishing the two harmonies provides a plausible defence. Spinoza's distinction seems to undermine effectively one of the most common criticisms of his system (the design argument for a God that acts from a free will).

Spinoza's use of two different conceptions of harmony raises two companion questions. How many concepts of harmony are there? And can we show that some of those concepts are related in interesting ways, such that accepting one conception entails, is compatible with, or is incompatible with another conception?

In the next section, I focus on the first question. I lay out eighteen different meanings of 'harmony' in the seventeenth and eighteenth centuries, and put forward evidence that many of these were in common use. I unearth positions and connections that challenge Spinoza's resolution of the *prima facie* problem, and focus on three challenges to Spinoza's metaphysical and ethical claims.

New Problems

We have seen that Spinoza introduces a distinction between one sort of harmony (*harmonia*) and another sort (*concordia*) but other conceptions abound. The philosophical uses of the term 'harmony' in the seventeenth and eighteenth centuries can be roughly grouped into five categories and eighteen core uses. These uses are neither exclusive nor exhaustive. Some philosophers use more than one of these notions of harmony without contradiction. Perhaps some meanings cannot be held together consistently, and perhaps some one use entails another, but I will not focus so much on the conceptual space as identify these uses and show how they open up or close off argumentative possibilities. As we will see repeatedly, it may not always be possible to distinguish whether a particular use falls cleanly into one of these meanings rather than another. Indeed, some of the uses I mention are merely possible readings of certain difficult to parse claims.

Consistency Uses

1. compossibility
   A and B are in harmony iff the existence of A is compossible with the existence of B
2. consistency
   A and B are in harmony iff there is no contradiction in both A being true and B being true
3. coordination

A and B are in harmony iff the actions, states, or properties of A correspond to or reflect the actions, states, or properties of B

I distinguish compossibility and consistency to note that harmony could be a claim about two possibly existent individuals or a claim about two statements, propositions, or facts. Coordination is suggested by Leibniz's pre-established harmony (1989: 78, 87–8), but Leibniz has at least one other use as well, discussed below.

Harmony as consistency appears most frequently in the following context. In the late seventeenth century, there were many popular 'Harmony of the Gospels' books published. These reordered passages from the Christian New Testament books of Matthew, Mark, and Luke (and sometimes John) to show their consistency in how they depicted historical events. Examples include John Eliot's *The Harmony of the Gospels* (1678); John Le Cler's *The Harmony of the Evangelists* (1701); William Whiston's *A Short View of the chronology of the Old Testament: and of the harmony of the four evangelists* (1702); and James Macknight's *A Harmony of the Four Gospels* (1763). Matthew Pilkington's *The Evangelical History and Harmony* (1747: x–xx) tracks 104 of these harmonies, starting with Tatian in the second century and increasing in number in 1537 (presumably due to the increase in Bibles published in the vernacular, which led to more laypersons reading the Bible and asking questions about its consistency). Philosophical theologians such as William Whiston published these harmonies to show that there was no contradiction between the biblical texts. These harmonisers may have wanted *more* than mere consistency. For instance, if things are harmonious then perhaps they are *beneficial* or *beautiful*. But at the very least, it would be necessary to posit consistency.

Commonality Uses

4. identity
   A and B are in harmony iff A is B
5. overlap
   A and B are in harmony iff A and B share a common part
6. medium
   A and B are in harmony iff there is some C such that C participates in the nature of A, and C participates in the nature of B
7. similarity
   A and B are in harmony iff A is sufficiently similar to B
8. unity in diversity
   A is harmonious iff A exhibits a high degree of similarity or identity among its parts, regions, states, or stages and also a high degree of diversity or variety among its parts, regions, states, or stages

The second cluster of uses focuses on harmony as identity, similarity, or some other way in which two things can be alike or have something in common.

Spinoza's positive use fits in this group, but where exactly it fits isn't settled. Most of Part IV of the *Ethics* is grounded in Spinoza's account that we care about others either because strictly speaking any being with a rational nature is identical to me *or* because other beings with rational natures are similar to me *or* because my rational nature is a part of me and it is also a part of you and therefore there is overlap between us. Michael Della Rocca (1994) and Jonathan Bennett (2001: I.217–23) have each argued that the different interpretations of Spinoza's claim are required by different arguments or texts and that there is therefore a deep and difficult-to-resolve tension in Spinoza's thought. I won't now settle this but merely flag it, because the issues that I address apply on any of these readings. While I will typically say that Spinoza's positive notion of harmony is one of similarity, the key is that there is a metaphysical basis in humans' rational nature, not in the imagination, and that this serves as the bedrock of Spinoza's ideal society.

Anne Conway in *The Principles of the Most Ancient and Modern Philosophy* has a similar ambiguity. In writing that members of the same species love each other and that there is love for all things based on the common substance, Conway (1996 [1690]: ch. 7 s.3) claims that 'This comes not only from the unity of their nature but also because of their remarkable similarity to each other.' Conway claims that it is both the unity of their nature (which seems to be a claim of identity) *and* similarity that grounds the love that members of the same species have for each other. She also makes a claim that Spinoza does not seem to endorse, which is that having a common substance with other beings can serve as a ground for loving every existing thing. Furthermore, for Conway, similarity as harmony is the basis for judgements of goodness.

What is the 'medium' reading? The sixteenth-century philosopher Jean Bodin argues that harmony in bodies involves an intermediate being that participates in the nature of two other things (Blair 1997: 134). Musical and civil harmony also require a middle term. There need be no overlap between A and B, but the third thing harmonises the other two.

Leibniz endorses the 'unity in diversity' understanding of harmony most clearly in the *Confessio Philosophi* of 1672. When prompted by the theologian, the philosopher defines harmony as 'Similarity in variety, that is, diversity compensated by identity' (Leibniz 2004: 29). This serves as the basis for many other claims, including that harmony is beauty and happiness and that God is the supreme model of harmony.

With the possible exception of Leibniz, no philosopher was more committed to harmony as an explanatory fact about reality than Jonathan Edwards. Excellency, the dominant concept in his early writings, is explained by harmony. The first section of *The Mind* begins with a challenge to the use of harmony as proportion (Edwards 1980 [1714]: 332). Equalities are the basis of beauty (Edwards 1980 [1714]: 335). He then reinforces the point about excellency being harmony being similarity.

> Excellency consists in the similarness of one being to another – not merely equality and proportion, but any kind of similarness. Thus similarness of direc-

tion: supposing many globes moving in right lines, it is more beautiful that they should move all the same way and according to the same direction, than if they moved disorderly, one one way and another another. This is an universal definition of excellency: The consent of being to being, or being's consent to entity. The more the consent is, and the more extensive, the greater is the excellency. (Edwards 1980 [1714]: 336)

Having defined excellency as harmony, he argues that greater similarity ('similarness') makes for greater harmony and excellence. One form of lesser similarity is proportion, which makes for a lesser harmony or excellence. Proportion may fall short in explanatory power, a problem Edwards seeks to correct through his account of similarity (Edwards 1980 [1714]: 332). But others did employ a proportionality account of harmony, which we shall consider next.

Proportionality Uses

9. proportionality of parts
   A and B are in harmony iff the parts of A are in a ratio of X:Y and the parts of B are in a ratio of X:Y
10. divine proportionality
    A is harmonious iff the parts of A are in a ratio that appears in the divine will or divine nature
11. universal proportionality
    A is harmonious iff the parts of A are in a ratio that appears frequently or fundamentally in nature (the universe)

Proportionality of parts could refer to physical parts, but doesn't always do so. Consider Henry More (1692: 215): 'For *Righteousness* is nothing else but an harmony of the lower parts of a mans Soul with the upper, of the Affections with Reason' (see also More 1653: 202–3). The 'parts' of the soul should probably not be considered as physical, extended, or compositional. John Locke (1824 [1690]: 143.4) considers a company of soldiers (or other people) and its component bodies: 'Besides, one angry body discomposes the whole company, and the harmony ceases upon any such jarring.'

The proportionality may exhibit itself in a manner that is modelled on the divine will or divine nature and receives its excellency for that reason. Returning to Edwards (1980 [1714]: s.1), 'the sweet harmony between the various parts of the universe is only an image of mutual love'. Mutual love is excellent, and its excellency comes from mirroring the divine nature. For others, the proportionality isn't modelled on the divine nature; it could reflect a structural feature of the universe itself. We saw such an account from John Heydon earlier.

Spinoza, we know, is critical of pleasingness accounts of harmony, which often appeal to proportion. He is not alone in this criticism. Nicholas Malebranche (1700: 3.10, 5.11) also makes it at multiple points in *The Search after Truth*. After discussing advances in astronomy, he notes that astronomers have held to an 'imaginary Harmony' about the regular motions of the planets. He also refers to the Pythagoreans' belief 'That the Heavens by their Regular Motions, made a wonderful Melody, which Men could not hear by reason of their being us'd to it.' 'But I only bring this particular Opinion of the Harmonical Proportion between the Distances and Motions of the Planets, to show that the Mind is fond of Proportions, and that it often imagines them where they are not' (Malebranche 1700: 3.10).

Agreement Uses

12. volitional agreement
    A and B are in harmony iff A is a goal of X and B is a goal of Y and the achievement of A is compatible with the achievement of B (where either X is Y or X is not Y)
13. divine volitional agreement
    A and B are in harmony iff A is a volition of God and B is a volition of God and the achievement of A is compatible with the achievement of B
14. divine-human volitional agreement
    A and B are in harmony iff A is a volition of God and B is a volition of a finite being and the achievement of A is compatible with the achievement of B

Another understanding of harmony is that one's volitions, decrees, or goals are compatible with each other or dovetail into some further end. Yet again, Edwards exhibits this view well (but this time in a later, unpublished work):

> God decrees all things harmoniously and in excellent order; one decree harmonizes with another, and there is such a relation between all the decrees as makes the most excellent order. Thus God decrees rain in drought because he decrees the earnest prayers of his people; or thus, he decrees the prayers of his people because he decrees rain. (Edwards 1994 [1722]: s.29)

In viewing harmony in this way, one person's volitions or ends could be compatible, or two different beings' volitions or ends could unite.

Pleasingness Uses

15. sensitive pleasingness
    A is harmonious iff A is pleasing to our senses
16. rational pleasingness
    A is harmonious iff A is pleasing to our reason
17. natural pleasingness
    A is harmonious iff A is pleasing to our nature
18. divine pleasingness
    A is harmonious iff A is pleasing to God

Harmony can be used to emphasise that something is pleasing to the senses, as Spinoza critically discusses in the appendix to Part I of the *Ethics*. While this is common, it could also be that something is pleasing to our reason, as More (1681: 194) states: 'Which coincidency of things to my reason is very harmonious . . .' In another work by More (1668: 326), a character (Philotheus) thinks 'Vice and Wickedness to my sense seeming so harshly repugnant to humane Nature, and Vertue [sic] and Righteousness so harmoniously agreeable thereunto'. So we could consider whether something is pleasing to my nature, without consideration of which aspect or part of my nature. We could also theologise the pleasingness and declare that what is harmonious is what is pleasing to God.

Francis Hutcheson (2008 [1738]: 1.1.14) has a robust account of harmony. After arguing that 'the Ideas of Beauty and Harmony, like other sensible Ideas, are necessarily pleasant to us', he examines what qualities occasion these ideas. Beauty requires a mind to sense the beauty, and beauty is finding the object 'agreeable to the Sense of Men', although animals may find something else agreeable due to their differing senses (Hutcheson 2008: 1.2.1). These qualities of objects that humans find agreeable are uniformity with variety, increased variety with equal uniformity, greater uniformity amidst equal variety, and a compound ratio (Hutcheson 2008 [1738]: 1.2.1; 2006 [1742]: 2.1.5). Hutcheson, probably aware of what music theorists were doing, was interested in laying out the specific features of objects or sounds that give rise to pleasingness in us. Put differently, working within the confines of Spinoza's negative characterisation of harmony as pleasingness, Hutcheson argues for a Leibnizian notion of unity with diversity. (Recall: 'Similarity in variety, that is, diversity compensated by identity' [Leibniz 2004 [1672]: 29].)

What Now of Harmony? What Now of Similarity?

Having examined some of the many things that harmony could mean in the period, we can now better understand what Spinoza is doing and how his talk about harmony connects to his other important commitments. Remember that Spinoza objected to harmony (*harmonia*, pleasingness) on the grounds that it did not describe real features of the universe but only features of our imagination which are grounded in our physical constitution. Because God does not share our physical constitution and does not know things through the imagination, it is inappropriate to ascribe this sort of harmony to God. In addition, there is a great deal of variety from person to person in what is found pleasing, which raises a concern about scepticism. However, when harmony is understood as similarity or unity through reason, Spinoza supports it.

Having now looked at many more conceptions of harmony than pleasingness and similarity, we can evaluate anew Spinoza's challenge to harmony. Making this distinction blocked a possible criticism of Spinoza's philosophy – that it contradicted itself in its reliance on and dismissal of harmony. But now we may ask, does Spinoza's challenge to harmony as pleasingness also challenge other conceptions, or might they be incorporated into his positive account?

Of the eighteen different conceptions of harmony discussed above, some are compatible with Spinoza's preferred positive use of harmony. Compossibility, consistency, and perhaps coordination are compatible with Spinoza's preferred similarity view of harmony. Because Spinoza is only interested in harmony in this world, any actual case of similarity-as-harmony will also be a case of compossibility (and truths about these cases will be examples of consistency).[3] He can also accommodate some proportionality conceptions of harmony into his positive account because if one thing's parts are in the same ratio as another thing's parts, then this is an instance of similarity. Spinoza's positive use of harmony (*concordia*) as similarity could collapse into harmony understood as identity if, as some have argued, he holds an account in which identity reduces to similarity, but this of course becomes complicated by issues beyond the scope of this chapter, including whether Spinoza holds to the identity of mind and body.

Spinoza's arguments against harmony clearly target what I have called 'sensitive pleasingness' and 'divine pleasingness'. To deal with those like Henry More (1681: 94; 1668: 326) who say that something is harmonious 'to my reason' and 'to humane Nature', Spinoza clearly intends to reclassify their appeal to reason as an appeal to imagination, as when he says that they have 'accepted affectations of the imagination as things' and that they don't reveal the world as it is but 'only the constitution of the imagination' (E IApp.).

Two potentially thorny conceptions of harmony are ones that Spinoza would plausibly have known about. First, the volitional conceptions of harmony are objectionable to Spinoza if they assume a *free* or *autonomous* will, but otherwise it is consistent with Spinoza's core commitments that there be 'harmonious' dovetailing acts that follow from the nature of things. Indeed, this might be an accurate account of the ideal society for Spinoza. Second, Leibniz's 'unity in diversity' conception of harmony is similar to Spinoza's understanding of God's being infinite. For Spinoza, 'By God I understand a being absolutely infinite, i.e., a substance consisting of an infinity of attributes, of which each one expresses an eternal and infinite essence' (E ID6), which is why '*From the necessity of the divine nature there must follow infinitely many things in infinitely many modes. . .*' (E IP16). This is clearly a positive feature of God, but Spinoza doesn't call it harmony. He reserves *harmonia* for approbation and *concordia* for similarity, so he might resist 'harmony' here, but I see no good reason why Spinoza cannot allow at least this one kind of unity in diversity (or better, diversity from unity) as a sort of positive harmony. It would not obviously reintroduce the forbidden move from the orderliness of the universe to the existence of a creator God with a free will, and it makes no clear appeal to imagination, so Spinoza's understanding of the divine nature could fit what others would call harmony.

Looking at harmony across the period can help us better understand the scope and the effectiveness of Spinoza's challenges to harmony. Having argued that the *prima facie* conflict between his positive and negative uses of harmony can be overcome, an examination of harmony reveals that Spinoza's two uses of harmony are just the starting point for the contentious concept cluster. Spinoza can accommodate most of these alternative uses, subsuming them into what he needs to make his negative argument and positive construal work.

A second way to use the exploration of harmony is to find new connections and possibilities that Spinoza did not recognise or could not foresee. One such connection is that Conway shares Spinoza's account of harmony as unity or similarity, despite Conway's critical attitude towards Spinoza, whom she thought a materialist pantheist (1996 [1690]: 9.3–9.5, 64–5). Despite their similar treatment of harmony and their commitment to the extended world being a single substance, they diverge in how they generate (explicitly from their statements about the similarity or unity of natures within a species) very different views of how animals can or should be treated.

According to Spinoza, the dog, the tree, and I are all modes of one substance. However, even though lower animals have sensations like we do (E IIIP57S), we can 'use them at our pleasure . . . for they do not agree in nature with us' (E IVP37S1). Conway raises the point in the *Principles* (but not in her response to Spinoza) that if all things share a common substance, this can ground ethical relations. 'Now, the basis of all love or desire, which brings one thing to another, is that they are of one nature and substance, or that they are like each other or of one mind, or that one has its being from another' (1996: 7.3, 46). Members of the same species love each other because they share a nature. Here Spinoza agrees. But even if there is not a unity of nature, there is still similarity across natures. Beyond this, all finite things share the same (single) substance, which can serve as the ground for the love of all things.[4] If we follow Conway in thinking that the unity of substance is sufficient to ground ethical relations or that there can be similarity across species – which Spinoza should also concede given that humans and non-human animals both have sensations (E IIIP57S) – then we could provide an environmental and animal ethic that is much more inclusive than that which Spinoza officially endorses. Conway and Spinoza agree that there is a difference in nature between us and other animals. The question is whether there is also sufficient similarity to ground ethical relations. Conway claims there is; Spinoza claims there isn't.

As critics since at least Margaret Wilson (1999) have pointed out, Spinoza's argument rests on a difficult-to-defend claim that animals and humans are different in nature – so different that we can use them 'at our pleasure'. A partial defence of Spinoza's argument by John Grey attempts to show that it is valid but not sound. Grey suggests that the least 'Spinozistic' part of the argument is Spinoza's 'doctrine of human nature' (2013: 81). Examining Spinoza's accounts of harmony suggests that his metaphysical claims about the similarity between humans is closely tied to another aspect of his ethical theory: that harmony is similarity, which is crucial to the development of his account of human society. From Conway, we can see how a drastically different account of how humans should treat non-human animals can be grounded on the same notion of harmony that Spinoza adopts in his ethical theory.

Spinoza, then, can be saved from the *prima facie* charge of inconsistency in his discussions of harmony, but for Spinozists to develop an animal ethic that is more humane than Spinoza's, they would do well to see how Conway starts from many of the same metaphysical and ethical premises but reaches a significantly different conclusion.[5]

Notes

1. I am using these professional labels anachronistically to emphasise the diversity of approaches and not because they conceived themselves as pursuing unrelated fields.
2. 'Those which move the sense through the nose, they call pleasant-smelling or stinking; through the tongue, sweet or bitter, tasty or tasteless; through touch, hard or soft, rough or smooth, etc.; and finally, those which move the ears are said to produce, noise, sound or harmony' (E IApp.).
3. If, as some have argued, the actual world is the only possible world for Spinoza (on the basis of passages that include E IP29 and E IP33), and a full account of any possible being includes its relations to everything else in that world, then we can further state that (without qualification) any case of similarity is a case of compossibility for Spinoza.
4. A limitation of the approach I am taking here is that Conway thinks God is a different substance than the extended, spiritual world. This is a two-substance view (with both substances being spiritual). Spinoza admits only one substance. While a significant difference, my argument here focuses only on the grounding of ethical relations among beings extant in the world. The two philosophers agree that there is a single substance that includes (in some sense) all finite beings, which are not themselves separate substances.
5. I am thankful to the participants at the 'Spinoza and Proportion' conference at the University of Aberdeen and the faculty at Mississippi State University for the opportunity to present and receive feedback on this work.

5

Ratio as the Basis of Spinoza's Concept of Equality

Beth Lord

What, for Spinoza, would constitute an equal society? I would like to address this question not on the basis of what Spinoza appears to say about equality in his political texts, nor on the basis of the egalitarianism that others have ascribed to him,[1] but on the basis of his own concept of equality. In this chapter I will develop the view that, for Spinoza, the concept of equality derives from Euclid, and is deeply connected to the complex concept of *ratio*.

As the chapters in this volume make clear, *ratio* is central to Spinoza's philosophy. In Latin, *ratio* has at least three meanings. First, and most obviously, *ratio* means reason: reason is the second kind of knowledge, through which we understand things adequately and truly. *Ratio* can also mean the reason why something exists or acts as it does: its causal or explanatory story. Every individual has a *ratio* that fully explains it, the full reason or rationale for that thing being what it is. Our own reasoning involves understanding the causes of things, or the reasons why they are as they are. When we understand a thing through *ratio*, we understand the *ratio* of that thing.[2]

Secondly, *ratio* can mean relation. We are constantly in relation to the other people and things that constitute our world and that we encounter in experience. Indeed, one of Spinoza's first claims about finite modes is that they necessarily relate to one another (E IP28). No finite, durational thing can exist or cause effects unless it is determined to exist and cause effects by another finite, durational thing, and so on, to infinity. A finite body and a finite mind exist, act, and think by virtue of their relations with other finite bodies and minds. The more relations a body enters into, the more ideas its mind – which enters into relations with the ideas of those things – is capable of perceiving. The highly composite and affective human body has a high degree of relatability, both internally, in terms of the interrelation of its constituent parts, and externally, in terms of its relations to other things. As our bodies are constantly exchanging affects with other bodies, and our minds are constantly exchanging ideas with other minds, we are constituted by these relations.[3]

Thirdly, *ratio* can mean a mathematical relation or proportion, as when Spinoza says that a body is constituted by a ratio of motion and rest. A composite body's characteristic ratio constitutes its identity through change, its distinctiveness from other bodies, and its capacity either to combine with those bodies in larger composites, or to damage and destroy other bodies. The body's ratio indicates the 'certain fixed manner', or pattern, according to which the body's parts communicate motion (E IIA2″D). Similarly, the whole of physical nature has a constitutive ratio of motion and rest that is preserved amid all its variations. In this sense, we can say that the physical universe has a characteristic proportion, or harmony, even if we cannot say that it is harmonious in a teleological or anthropocentric sense.[4]

Bodies are governed by ratios in the sense of mathematical proportions. Minds can also be said to be governed by ratios, in a different sense. The finite mind, insofar as it endures, is the idea of the body, constituted by ideas of its body's parts, processes, and relations. Most of these ideas are inadequate, but some are adequate, and we can gain more adequate ideas as we think more rationally. Spinoza is clear that a person's mind cannot consist exclusively of inadequate or adequate ideas: everyone has some of each. At a given moment, then, a proportion of the human mind is constituted by adequate ideas, while another proportion is constituted by inadequate ideas. Minds with a greater proportion of adequate ideas are more intellectually active, perfect, and blessed; it is this proportion of the mind that Spinoza calls its 'eternal part' (E VP39–40; LeBuffe 2010). The finite mind can thus be expressed as a ratio of adequate to inadequate ideas. This ratio itself can change as we increase our reasoning, but it consistently corresponds to one body. We might say that the mind's variable ratio is the idea of the body's fixed ratio, or indeed, that the mind's variable ratio and the body's fixed ratio are one and the same thing.

The three senses of *ratio* – reason, relation, and proportion – are complexly intertwined. Our relations with other things determine, and are determined by, the ratio of motion and rest that constitutes our body's form. Our relational interactions with other things are crucial in determining what proportion of the mind is adequate. The more positive relations a body experiences, the more reasoning its mind is likely to develop, whereas affections, deriving from relations, impede the mind's reasoning. The proportion of the mind that is constituted by adequate ideas reflects the extent to which that mind reasons. Ethically we aim to have positive relations, maintain our bodily ratio, improve our mental ratio of adequate to inadequate ideas, and improve our rational understanding of the reasons why things are as they are. Finally, these bodily and mental ratios provide the reasons why things act as they do. *Ratio* is a complex concept that is central to Spinoza's metaphysics, his doctrines of body and mind, and his ethical theory, and I will argue here that *ratio* is the basis of Spinoza's understanding of equality. Specifically, I will argue that Spinoza understands equality not as the equivalence of interchangeable units, but as the equivalence of proportions. That is, for Spinoza, equality is fundamentally geometrical equality. Geometrical equality is the basis of the civil state and the other forms of equality the best state can feature.

Geometrical Equality

For Spinoza, the concept of equality is first and foremost geometrical. In the *Ethics*, as in the earlier *Treatise on the Emendation of the Intellect*, *Short Treatise*, and *Principles of Cartesian Philosophy*, the term 'equal' is typically found in references to geometrical principles.[5] These principles come from Euclid's *Elements*, the text that gives Spinoza the model for the geometrical method and a paradigm for adequate thinking.[6] Spinoza claims that mathematics, understood as geometry, shows us the 'standard of truth', for it is concerned 'only with the essences and properties of figures' (E IApp.). The objective essences of figures are true ideas, and the complete definition of a figure is the explanation of its essence (TIE 33–6, 93–6).[7] To think through a geometrical definition, then, is to think truly: it is to have an adequate idea of the figure from which its properties may be deduced. Geometry is paradigmatic of adequate thinking, and Euclid's principles are examples of how we use good definitions to understand the essences of things and acquire adequate knowledge (see, e.g., TIE 72).

Equality is not explicitly defined in Euclid's *Elements*, but its proper uses are set out in the five axioms (also known as common notions) of Book I. From these axioms we understand that between magnitudes of the same kind ('magnitudes' being geometrical entities that take up space such as lines, angles, or figures), equality is a relation of equivalence involving reflexivity (for each x, $x = x$); symmetry (if $x = y$, then $y = x$); and transitivity (if $x = y$ and $y = z$, then $x = z$). Equality does not always involve identity, for equivalence may be posited between things of the same kind that differ. While all magnitudes that coincide (that is, that can be spatially superposed) are equal, not all equal magnitudes coincide: magnitudes may differ in spatial form and position and yet be equal. For Euclid, then, equality may be posited of magnitudes in three cases: the case of numerical and spatial coincidence (e.g. a single triangle equal to itself); the case of numerical difference and spatial coincidence (e.g. two triangles that can be superposed); and the case of numerical difference and spatial non-coincidence (e.g. a triangle and a square having the same area). The first case is equality as identity, but in the second and third cases, equality incorporates sameness and difference. Two magnitudes that differ in one respect are said to be the same in another respect. This is what we will call geometrical equality.

Both Descartes and Spinoza draw on geometrical equality in a frequently cited example: Euclid's principle that a triangle's three angles equal two right angles. In any triangle, the three angles differ from two right angles in several respects, but have the same value. Here we have an assertion of geometrical equality, incorporating sameness and difference. For both Descartes and Spinoza, this proposition is the geometrical model *par excellence* of the deduction of the properties of a thing from the true idea of its essence.[8] In Descartes' fifth Meditation, the meditator's knowledge that the triangle's three angles equal two right angles is found to be grounded in the mind-independent essence of the triangle. This leads to Descartes' objective proof of God's existence, for in just the same way, the knowledge that God's properties include existence is grounded in the mind-independent essence of God. 'It is quite evident that existence can no more be separated from the essence of God than the fact that its three angles equal two right angles can be separated from the essence of a triangle' (Descartes 1996: 46 [AT VII, 66]; cf. Spinoza's commentary at PPC IP5S).

Spinoza makes a similar move in arguing that 'infinitely many things in infinitely many modes' follow from God's essence (E IP16). First Spinoza appeals to the geometrical process of deducing properties from definitions:

> This proposition must be plain to anyone, provided he attends to the fact that the intellect infers from the given definition of any thing a number of properties that really do follow necessarily from it (that is, from the essence of the thing); and that it infers more properties the more the definition of the thing expresses reality, that is, the more reality the essence of the defined thing involves. (E IP16Dem.)

The definition of God, like the definition of a geometrical figure, explains its essence. From the definition of God ('a substance consisting of an infinity of attributes, of which each one expresses an eternal and infinite essence' [E ID6]) follows an infinite number of properties, just as from the definition of a triangle follow the properties pertaining to its nature. Accordingly, the triangle analogy appears in the scholium to the next proposition:

> I think I have shown clearly enough (see P16) that from God's supreme power, *or* infinite nature, infinitely many things in infinitely many modes, that is, all things, have necessarily flowed, or always follow, by the same necessity and in the same way as from the nature of a triangle it follows, from eternity and to eternity, that its three angles are equal to two right angles. (E IP17S)

To deny that a triangle's three angles equal two right angles is to have a confused idea of what a triangle is. Similarly, to deny that all things follow necessarily from God's nature is to have a confused idea of God.

Geometrical equality is significant in this analogy. Euclid's principle that *a triangle's three angles equal two right angles* is an assertion of equivalence between two sets of angles that are not identical and do not coincide. The three angles cannot be spatially superposed on the two right angles that they equal; furthermore, the three angles are variable in value (provided that they add up to 180 degrees), whereas the two right angles are fixed. What follows from the triangle's essence is a relation of *sameness and difference*: three variable angles that are both the same as, and different from, two invariant angles. On this account, what makes the triangle what it is, and what defines it essentially, is not the relation of identity expressed through the phrase 'a triangle is a three-sided figure', but a relation of geometrical equality that incorporates sameness and difference. Geometry makes it possible for us to maintain sameness and difference together in our thinking without being obliged to oppose them or synthesise them. Sameness and difference are neither taken in isolation, nor collapsed into one another, but are reciprocally related to one another.⁹ It is these reciprocal relations, rather than statements of identity, that define things essentially.

By linking the triangle to God, Spinoza leads us to consider how God's essence is defined by geometrical equality and the reciprocal relation of sameness and difference. By maintaining sameness and difference together in our thinking, we might better understand how God's attributes are *the same* as one another, and *distinct* from one another at the same time.¹⁰ We might escape the difficult problem of the identity of the attributes by saying that the attributes are not identical, but equal. Spinoza seems to support this line of thought when he states that 'God's power of thinking is equal to his actual power of acting' (E IIP7C; there is a similar use of 'equal' at IIIP28Dem.). This statement is made in the corollary to the parallelism proposition, in which Spinoza states that the order and connection of ideas *is the same as* the order and connection of things, although the two are causally and conceptually *distinct*. Spinoza is saying that we must understand the two streams of causality to be both the same and different – a complex notion summed up by the word *equal* when it is taken to mean geometrical equality.¹¹ On this reading, God's essence would consist in the geometrical equality of non-identical and non-coinciding attributes, just as the triangle's essence consists in the geometrical equality of non-identical and non-coinciding sets of angles. 'Equal' is shorthand for the complex way in which God is at once unity and plurality, expressing both self-sameness and self-difference.

The equality of the attributes is, however, a discussion for another time. Here we need to understand how geometrical equality, incorporating sameness and difference, can give way to a model of the equal society. This is where we come back to ratio. What will allow us to get from geometrical equality to the equality of people in a state is the notion that physical bodies, through their equal ratios, involve geometrical equality.

The Geometrical Equality of Bodies

In this section I will set out how the parts of a physical body are geometrically equal. I will then discuss how a physical body may be considered geometrically equal to other physical bodies.

For Spinoza, bodies are not distinct substances but modes of one extended substance distinguished by different rates of motion. Every individual is understood both as a 'whole' composed of a union of parts, whose rate of motion differs from those of other wholes, and as a 'part' of a greater whole, whose rate of motion agrees with those of other parts (see E IIL7 and Ep. 32 to Oldenburg, 1665). The individual's difference from and agreement with others is based on its characteristic ratio of motion and rest, a term Spinoza never entirely clarifies. In the physical interlude between Propositions 13 and 14 of *Ethics* Part II, he claims that 'the simplest bodies' either move or are at rest, and that each one is capable of moving at a varying speed (E IIA1', A2', and A2"). The speed of each simple body is determined by the motion and rest of other bodies, meaning that (recalling E IP28) even the simplest bodies are defined in terms of their relations to other things. Composite bodies are defined additionally by the internal relations of their parts: a composite body is made up of a number of bodies that are physically constrained together and whose inter-determination consists in 'communicat[ing] their motions to each other in a certain fixed manner' (E IIA2″D). So long as its parts continue to determine one another in this way, the composite body is united, and composes one individual. The union that is produced by the particular way the parts determine one another constitutes the form of the individual, which distinguishes one body from another.[12] The individual's union remains constant as it gains and loses parts and as its parts alter direction and speed (E IIL4–6).

Crucially, the form of a body consists in its external relations to other bodies and the internal relations among its parts. The *ratio* (relationality) of its form is expressed by its *ratio* of motion and rest, which determines the identity of the composite body through change:

> If the parts composing an individual become greater or less, but in such a proportion [*proportione*] that they all keep the same ratio [*ratione*] of motion and rest to each other as before, then the individual will likewise retain its nature, as before, without any change of form. (E IIL5)

> [W]hat constitutes the form of the human body consists in this, that its parts communicate their motions to one another in a certain fixed proportion [*ratione*]. Therefore, things which bring it about that the parts of the human body preserve the same proportion [*ratio*] of motion and rest to one another, preserve the human body's form. (E IVP39Dem.)

The form of a body consists in its *ratione*, the 'certain fixed manner' in which the body's parts communicate motion to one another. It is not clear whether Spinoza believes this to be expressible as a precise mathematical ratio, as Matheron (1969: 39–41) suggests. In a footnote to the *Short Treatise* Spinoza says that the body is defined by a ratio, 'say e.g., of 1 to 3' (KV IIPref.n.1), but this may have been added by another author (see Gabbey 1996: 189). In the 'letter on the infinite' (Ep. 12 to Meyer, 1663), Spinoza argues that number is merely a way of imagining, as is the division of extended substance into parts: no numerical definition truly describes reality (see also E IP15S). Jacob Adler (2014) intriguingly argues that the 'ratio of motion and rest' indicates an imprecise equilibrium, stemming from Spinoza's attempt to mechanise the Renaissance doctrine of the proportionality of the four humours.[13] The ratio of motion and rest is the way the body (or perhaps a particular part of that body – the brain) coordinates the many and various relations among its moving parts and its many and various (and changing) relations to external bodies. In this sense, the ratio of motion and rest is the body's coherent and coordinating relationality, best expressed not by any one mathematical ratio but by proportion (another plausible translation of *ratione*).

Euclid defines as *proportional* magnitudes that have the same ratio.[14] Spinoza's ratios of motion and rest, being in the attribute of extension, are relations between magnitudes. According to Euclid, the proportionality of ratios of magnitudes is ascertained differently from the proportionality of ratios of numbers.[15] We establish that ratios of numbers (say, 1:2 and 3:6) are proportional by cross-multiplication. If the products are equal, then the numbers express the same ratio and are proportional (*Elements* VII, Def. 20). Spinoza refers to this procedure at E IIP40S2 when he uses the problem of the fourth proportional to illustrate the three kinds of knowledge (see also TIE 23–4 and KV II.1). By contrast, we establish that ratios of magnitudes (say, $x:y$ and $a:b$, where x and y are two sides of triangle P, and a and b are two sides of triangle Q) are proportional by using common multipliers to make them larger or smaller. If certain relations remain the same through these changes, then the magnitudes express the same ratio and are proportional.[16] Note that Spinoza's first mention of the ratio of motion and rest in the *Ethics* is at IIL5, where he writes of 'the parts composing an individual becom[ing] greater or less, but in such a proportion that they all keep the same ratio'. Spinoza explains how we ascertain the proportionality of a body's parts in the same way as Euclid does for magnitudes: if, as they increase or decrease in size, they maintain a certain relation, then the ratio that connects them remains the same and they are proportional. A body, for Spinoza, can be seen to be made up of parts that are geometrical magnitudes; it may even be that when he speaks of 'the simplest bodies' (E IIA2″) he means points and lines, rather than atoms or particles of motion.[17]

In any case, the body's parts maintain their proportionality as they increase or decrease in size. Alternatively, we may say that the ratios of the parts of a body are *equal* to each other. 'Equal' is the appropriate word because the ratios are both the same and different. Every individual body is distinguished by its characteristic ratio of motion and rest, and each *part* of a body is also an individual. Each part, then, has its own characteristic ratio that differs from the ratios of all the other parts, without which it would not be distinct from them. Yet in order to maintain the body's proportion and identity through change, those ratios must nevertheless be 'the same'. The body incorporates difference and sameness in respect of the ratios of its parts: each part has its distinctive ratio *and* has the same ratio as all the others (insofar as the body's proportion persists). Here we have a case of geometrical equality. The body's parts are geometrically equal, at once different and the same ($a:b = c:d = e:f = g:h \ldots$). It is crucial that they maintain both difference and sameness: each part must continue to exist as its distinctive ratio, which makes it what it is and allows it to fulfil its specific function, and the parts must continue to have the same ratio lest the whole body be destroyed. A body that strives to maintain its identity and 'preserve the same proportion of motion and rest' among its parts (E IVP39Dem.), then, strives to maintain *the geometrical equality of its parts*. The maintenance of geometrical equality is felt as cheerfulness, the affect in which all the body's parts are 'equally affected' by joy (E IIIP11S and IVP42Dem.).

A physical body is not only a whole in its own right, but also a part of larger wholes. The human body can join together with other bodies to form a larger, unified body that is capable of coordinated movement, such as a team of rowers.

The larger body has its own characteristic ratio, perhaps determined by the cox in a team of rowers, who strives to maintain the geometrical equality of the parts of the whole.[18] That is, the cox determines a coherent and coordinating relationality through which she and all the rowers, each with their own distinctive ratio, move proportionally so that all together maintain the same overall ratio. To the extent that a human body becomes a part of a whole like this, it is equal to the other human bodies in that whole. Although this is strictly *geometrical* equality, this point is striking: it seems to be an implication of Spinoza's physical theory that one human body is *equal* to another human body, insofar as they move together and cooperate as parts of a single relational and proportional whole.

Can the same be said of other human groupings capable of coordinated movement, such as the diverse crew members of a large ship or citizens of a state? Are the bodies in these groupings parts of a whole that has a single ratio of motion and rest? Spinoza appears to suggest so at E IID7 and IVP18S, leading some interpreters to argue strongly that they are. Others have denied that communities are 'individuals' in the sense set out in the physical digression.[19] Spinoza's strongest statement on this is the 'worm in the blood' letter (Ep. 32 to Oldenburg, 1665), in which he says that what constitutes parts and wholes is a matter of perspective. The worm that lives in the blood sees the objects around it as wholes, and does not see how they are 'restrained by the universal nature of the blood' nor how they are compelled to harmonise with one another. By analogy, a human being does not see how apparently disconnected objects in the galaxy are truly unified and forced, by the laws of nature, to work coherently. Spinoza stresses that 'by coherence of parts' he understands 'that the laws or nature of the one part so adapt themselves to the laws or nature of the other part that they are opposed to each other as little as possible'. Spinoza's description of the coherence of parts here is not as strict as that of the physical digression of the *Ethics*. Parts of a whole do not need to lie upon one another or move as one body; they simply need to adapt themselves to each other so that there is minimal opposition and maximal agreement between them. It seems to follow that anywhere this adaptation and agreement take place – that is, anywhere that coherent relationality obtains – a whole with a single ratio can be identified. Spinoza famously extends this principle to the whole universe:

> Now all the bodies in Nature can and should be conceived in the same way as we have here conceived the blood; for all bodies are surrounded by others and are reciprocally determined to exist and to act in a fixed and determinate way, the same ratio of motion to rest being preserved in them taken all together, that is, the universe as a whole. Hence it follows that every body, insofar as it exists as modified in a definite way, must be considered as a part of the whole universe, and as agreeing with the whole and cohering with the other parts. (Ep. 32 to Oldenburg, 1665)

This passage complements Spinoza's claim in the *Ethics* that the whole of extended nature is one infinitely composite individual 'whose parts, that is, all bodies, vary in infinite ways, without any change of the whole individual' (E IIL7S). All the bodies in the physical universe, it seems, relate coherently in a single proportional whole.[20] Nature has one unified ratio that is expressed by each individual body, even as each individual body expresses its own distinctive ratio. This suggests that every body in nature is geometrically equal to every other body.

This may appear a trivial point, given that what we are interested in is whether certain groups of bodies – specifically, communities of human beings – can be counted as physical individuals. Yet this most generic kind of geometrical equality is the starting point for Spinoza's political philosophy. Unlike Hobbes, Spinoza does not believe that human beings are equals in the state of nature, except in this very generic sense: as a part of the whole of nature, every person is geometrically equal to every other person, and to every animal, plant, and mineral in nature too.[21] In forming a civil state, people remain 'equal under nature' in this sense (hence Spinoza's claim that in a democracy, 'all remain equal as they had been previously, in the state of nature', TTP ch. 16/G III 195), but they additionally commit to constituting an exclusively human group with its own distinctive ratio or coherent relationality. In doing so, these people become 'equal under the civil state': already geometrically equal in the generic sense as bodies in nature, they become geometrically equal in a specific sense as bodies that constitute a human community. A human community may include one or more members that govern the coordinated movements of the others, but those members (like the cox in the team of rowers, or the brain in the body) remain geometrically equal to the other parts. Their geometrical equality means that each one agrees to contribute to the coherent relationality of the whole community while simultaneously striving to preserve his or her own physical ratio. That is, committing to geometrical equality in the civil state is the social contract itself.

The Equal Society

As we have seen, the foundation of the civil state involves a commitment to geometrical equality. This is simply the assertion that the distinctive ratio of each member of a state coheres with the ratio of the whole community: it is not political, social, legal, or economic equality. Nor is it equality of individuals' power to strive and thrive. Yet the notion of geometrical equality can help us to reflect on what an equal society looks like, for Spinoza.

The language Spinoza uses to describe the coherence of bodies in Ep. 32 – that of adaptation of natures, agreement, and harmony – is the same as the language he uses to discuss social and political communities. Spinoza's political philosophy, like Hobbes's, is founded on the metaphysics of parts and wholes. The social contract is an agreement, by those who are parts of the whole of nature, to become parts of a human whole working towards human goals. We become citizens through an agreement to cease to be sovereign wholes in our own right, and to become parts of a unified whole that is sovereign. Our capacity to become parts of such a whole is based on the agreement of our natures as human beings:

because human beings can agree in nature, they can work together for mutual utility (E IVP31, P35).

For Spinoza, 'agreement in nature' is a matter of degree. He suggests that things agree in nature to the extent that they have anything in common. This means that any two things within the same attribute (and thus, any two things in the physical universe) can agree in nature to some degree (see E IVP29Dem.). Things agree in nature more, the more they have in common, and the less harm they can do to one another through the contrary aspects of their natures (see E IVP29–P35). Things with a high degree of agreement are better able to join together as physical wholes, because their natures are more adaptable and there is less potential for opposition. A human community is based on the fact that human beings can agree to a high degree (although they do not always do so). A community can thus be understood as a physical whole made up of parts whose natures adapt themselves to one another's to some extent, typically through coordination and governance by one or more powers.

Spinoza sets out the concept of community as a *physical* whole in the scholium to IVP18:

> There are ... many things outside us which are useful to us, and on that account to be sought.
> Of these, we can think of none more excellent than those which agree entirely with our nature. For if, for example, two individuals of entirely the same nature are joined to one another, they compose an individual twice as powerful as each one. To man, then, there is nothing more useful than man. Man, I say, can wish for nothing more helpful to the preservation of his being than that all should so agree in all things that the minds and bodies of all would compose, as it were, one mind and one body; that all should strive together, as far as they can, to preserve their being; and that all together should seek for themselves the common advantage of all.

Spinoza alludes more frequently to the Hobbesian notion that a community is a single physical body in the *Political Treatise*, where he also says that a community is guided *una veluti mente*, 'as if by one mind' (see, e.g., TP 2.15–16, 3.2, and 3.5). In the *Ethics* passage, he similarly presents the notion that a community has 'one mind and one body' as a heuristic condition to be wished for. We wish for a high level of agreement among human beings such that a human community resembles a unified physical body with a single mind striving to preserve its being. Of course, human beings frequently have high levels of disagreement, and most communities do not operate with the unified striving and action of a single physical body and mind. But at its very best, a human community is a whole whose parts operate coherently with a single coordinating power, just as they do in a physical body. That suggests that the best human community is one in which each individual part maintains his or her distinctive ratio and is coordinated into the coherent relationality of a unified physical whole. In other words, the best human community is one whose parts are geometrically equal.

The geometrical equality of bodies in a community means that everyone adapts to one another's natures and relates to all the others in a coherent manner, maintaining the society's overall ratio as the community grows, shrinks, and changes. At the same time, all the members of the society maintain their distinctive ratios, pursuing their own interests and working to their own strengths. Such a society would be highly functional and stable through change. Stability through change – that is, security – is one of the purposes of the state, for Spinoza: 'the best state is one where men live together in harmony and where the laws are preserved unbroken' (TP 5.2). A geometrically equal state fulfils this purpose in that its members have high levels of agreement and maintain the coherent relationality of the community, preserving the state's form and identity, and the laws by which its members relate to one another.

The other purpose of the state is the ethical freedom of the people who live in it (TTP ch. 20/G III 240–1).[22] A good state strives to develop people's capacities for rational thinking, virtue, and autonomous action. A geometrically equal state fulfils this purpose too, for in order to keep everyone's ratios in agreement, it must foster positive relations between its members and minimise dangerous passions. A geometrically equal state strives to maintain its overall ratio, and to maintain the particular ratios of its constituent members. It will therefore aim to minimise strong passions and to encourage positive interactions and relationships that are 'good' for 'the preservation of the proportion of motion and rest the human body's parts have to one another' (E IVP39). To preserve the ratios of parts and whole, the state (or its governors) must ensure that everyone can meet their physical needs – not only for food, drink, and shelter, but also for 'decoration, music, sports, the theater' (E IVP45S) and other goods that encourage cheerfulness (the feeling of the body's proportionality). Only if they maintain their own coherent relationality can citizens continue to coordinate their movements effectively, flourish physically, and act from their own nature, enabling them to become more rational, virtuous, and free.

A geometrically equal society can give rise to other kinds of equality. Economic equality, achieved through the equitable distribution of goods and the encouragement of mutual aid, enables everyone to get what they need and enhances societal agreement and cohesion (see Lord 2016). A geometrically equal society that strives to remain so is likely to feature high levels of economic equality too. Note, however, that this society need not feature social or political equality. A community may be stable, functional, and harmonious, with a high level of cooperation, mutual aid, and flourishing, without its members being socially or politically equal.[23] What matters to Spinoza is not that everyone has an equal right to vote or is equally entitled to moral respect, but that everyone has an equal opportunity to flourish. A geometrically equal society is not one in which every individual is taken to be of equivalent *value* to every other, for the value of a person consists in their rational development, which varies according to their power (Lord 2014). In a geometrically equal society, we strive that everyone may flourish as much as they can, meaning that some will be much more rational than others. This is, then, a society in which different individuals, each with their own ratio, may be differently valued (according to their different levels of power and reasoning, concomitant with their different physical and mental ratios), but are equitably enabled to flourish in order to maintain the relationality of the whole. In Spinoza's 'best' society individuals are not understood as equally valued, interchangeable units (1 = 1 = 1 . . .). Instead, they are understood as differently powered and differently valued but geometrically equal ratios ($a:b = c:d = x:y$ . . .), whose differences are as important to uphold as their sameness.[24]

I hope to have shown here that Spinoza does have a concept of the equal society. It may not be the one that we are most familiar with, for it is not premised upon the moral equivalence and political equality of individuals in the democratic state. Instead, it is based on geometry, which gives rise to the kind of equality Spinoza thought most important: the equality of flourishing. This kind of equality is deeply bound up with the Latin term *ratio* in all its senses: ratios of motion and rest, the coherent relationality of the whole, the need for positive social and affective relations, and the development of reasoning. Spinoza's philosophy of *ratio* does not apply only to mathematics, physics, and metaphysics, but also to our thoughts about how to constitute workable social and political wholes of individuals whose equality consists in their simultaneous difference and sameness.[25]

Notes

1. I have approached the question from these perspectives in Lord (2016) and (2014).
2. On these senses of *ratio*, see LeBuffe's chapter in this volume.
3. On this sense of *ratio*, see Lloyd (1994) and Armstrong (2009).
4. On these themes, see the chapters by Ravven and Yenter in this volume.
5. Specifically, the principle that a triangle's three angles equal two right angles (PPC I Proleg. and IP5S; CM II.9; E IP17S, IIP49, and IVP57S, from Euclid's *Elements* I, Prop. 32), and that the lines drawn from the centre to the circumference of a circle are equal (TIE 95–6; PPC IP7S; CM II.10; E IP15S, from Euclid's *Elements* I, Def. 15). All references to Euclid's *Elements* are to Joyce (1998), on whose commentary I have relied substantially.
6. For an overview of Spinoza's use of the Euclidian method, see Steenbakkers (2009). See Viljanen (2011) on the significance of Spinoza's geometrical thinking to his metaphysics, and Barbaras (2007) on its affective and ethical character.
7. For a richer explanation see Viljanen (2011: ch. 1). For Spinoza, true understanding of nature proceeds from geometrical definitions and essences, not from measureable quantities, which are imaginary ways of dividing nature; see Ep. 12 to Meyer, 1663, and Manning (2012).
8. This proposition is also important for Hobbes. On its significance for early modern philosophy, see Viljanen (2011: 12–16, 43–4).
9. This is an example of what Balibar (1997: 10) calls 'the logic of the simultaneous rejection of abstract opposites' in Spinoza. Jaquet (2004: ch. 1) also draws out this relational aspect of Spinoza's concept of equality.
10. For a recent discussion of this problem and some solutions to it, see Schmidt (2009).
11. This is also suggested by Jaquet (2004: 34–5). Deleuze (1992: 69–70) and Macherey

(2011: ch. 3) also stress the equality of the attributes, but appear to have in mind lack of hierarchy rather than geometrical equality.

12. Cf. the *Short Treatise*: 'The differences among [bodies] result solely from the varying proportions of motion and rest, through which this is *so*, and not *so* – this is *this*, and not *that*' (KV IIPref.n.1).
13. On the body's equilibrium, see also Ravven's chapter in this volume.
14. 'Let magnitudes which have the same ratio be called proportional' (*Elements* V, Def. 6).
15. Matheron (1986) discusses this distinction and argues that both kinds of proportion are significant for Spinoza's understanding of bodies, but I must admit to finding his paper unintelligible.
16. 'Two ratios [of magnitudes] $w{:}x$ and $y{:}z$ [are] the same, written $w{:}x = y{:}z$, when for all numbers n and m it is the case that if nw is greater, equal, or less than mx, then ny is greater, equal, or less than mz, respectively' (*Elements* V, Def. 5).
17. There is, however, some evidence to the contrary in E IP15S. Since there is no real distinction between the parts of matter for Spinoza, the 'simplest bodies' cannot be atoms, leading some commentators to propose a 'wave' theory of matter (Bennett 1984: ch. 4; Viljanen 2011: 157–67). Spinoza claims that it is legitimate to distinguish parts 'modally' insofar as matter moves at different rates (E IP15S), but does not define motion; see Gabbey (1996), Manning (2012), and Peterman (2014) for discussion. It seems to me plausible that among the sources of Spinoza's physical theory is, alongside the Cartesian theory of motion, Euclid's doctrine of geometrical magnitudes. Andrea Sangiacomo has helpfully suggested that Spinoza may be drawing on Hobbes, who develops a theory of magnitudes as proportions in *De Corpore*.
18. In thinking through this example I have found Godfrey-Smith (2017: ch. 2) helpful.
19. That communities are individuals in Spinoza's sense is argued for most prominently by Matheron (1969), whose interpretation is the basis for much recent literature on Spinoza and the power of collective groups (or 'multitudes'; see, e.g., Montag 2005). For the opposing view, see, e.g., Rice (1990).
20. Oldenburg comments (Ep. 33 to Spinoza, 1665) that this doctrine appears to be incompatible with Spinoza's denial of order and harmony in nature (in E IApp.); for discussion of this problem, see Yenter's chapter in this volume.
21. See TTP ch. 16/G III 189–96, and Lord (2016) for further elaboration. Note that the geometrical equality of all beings – their coherence in one unified nature – is compatible with vast inequalities of power between different individuals.
22. For further discussion of how ethical freedom is achieved in the state, see, e.g., James (1996).
23. Spinoza's *Political Treatise* offers several models of functional societies that feature political inequality and social hierarchy.
24. On the importance of balancing sameness and difference in the state, see James (1996) and Lloyd (1994).
25. I would like to thank the many audiences who listened to earlier versions of this chapter and the individuals who offered questions and points that helped me to develop it. I would especially like to thank Keith Green for his detailed comments on an earlier draft.

6

Proportion as a Barometer of the Affective Life in Spinoza

Simon B. Duffy

As a catalyst for thinking about what an affective life might be and how such a life might be intimately bound up with its relations to its surroundings or environment, I propose in this chapter to present two different ways of thinking about individuality in Spinoza. I do so in order to draw out what is at stake in a double point of view of the degree of the power to act of a singular thing in Spinoza's *Ethics*. Sometimes this power to act seems to be fixed to a precisely determined degree, whereas sometimes it seems to admit a certain degree of variation. The problem of how to resolve this apparent contradiction has generated varying interpretations among scholars in Spinoza studies, including Martial Gueroult (1974), Gilles Deleuze (1992), Pierre Macherey (1995), and Charles Ramond (1995). The problem cannot be addressed in isolation from two other equally perplexing questions: whether the essence of a singular thing remains fixed or should be understood to be variable; and whether the *conatus*, as the expression of the power to act of a singular thing, should be understood to remain fixed or to be variable. These three questions are related because the three aspects of Spinoza's philosophy with which they deal – the power to act, modal essence, and *conatus* – are intimately intertwined, the interpretation of one having a bearing on the interpretation of the others. For a more precise understanding of these different ways of variation, and to render them compatible with each other, it is necessary to commence with the question of the variation of the essences of singular things. Certain interpreters, such as Gueroult, consider these essences to be situated between a 'minimum' and a 'maximum', and that it is only above or below this range that a rupture occurs in an individual's identity, resulting in a change of the individual's structure and therefore of its nature (Gueroult 1974: 351).

In *Quantité et qualité dans la philosophie de Spinoza*, by contrast, Charles Ramond argues that the essence of a singular thing is determined by a 'precise relation' of movement and rest, or by a *quantum* of the power to act of the *conatus*, and all variation is prohibited, since any augmentation or diminution of this power to act would create another individual. Despite this conclusion, Ramond believes that it leaves the concept of the human being 'absolutely incomprehensible, and moreover contrary to good sense' (Ramond 1995: 194). Ramond is here confronted with the apparent contradiction raised by our first question, which he wants to resolve by redefining one side of the contradiction. He maintains the concept of a fixed power to act of a singular thing, while denying that there is room for a margin of variation 'of' its power to act. He proposes to do this in the following way: 'if we want to be able to maintain a certain capacity of variation of the power to act of an individual who remains identical in its essence despite these variations, it will be necessary to separate in some way essence and power [*puissance*]' (Ramond 1995: 194). Ramond considers this possible only by changing our concept of the way in which the power to act varies. Ramond proposes to do this by making a distinction between changes 'of' the power to act and changes 'in' the power to act. By interpreting the *Ethics* according to the latter formulation, Ramond retains the concept of an unchanging power to act. 'There are therefore positive variations "in" the power to act of the body, and negative variations' (Ramond 1995: 193), while there are no variations 'of' the power to act itself. Ramond justifies this formulation by arguing that the 'variations "in" the power to act' occur within the theory of the passions and are therefore imaginary, that is, the power to act itself remains unchanged by these variations. 'This moreover', writes Ramond, 'does not surprise us, since such a distinction is at the base of the Spinozist theory of the passions' (1995: 193).[1] Despite the fact that Ramond offers a useful rubric to resolve the apparent contradiction, the two principal interpreters who bring to light exactly what is at stake in this question are Pierre Macherey and Gilles Deleuze.

Deleuze's interpretation of the dynamic changes in an individual's power to act diverges somewhat from that of Macherey. For Macherey, dynamic changes are incorporated by an individual according to the varying degree or proportion to which the active expression of its fixed power to act is inhibited or limited. For Deleuze, by contrast, an individual's power to act is open to 'metaphysical' or ontological changes. An individual for Deleuze is limited by the passive affections that it experiences in its interactions with other bodies, which, at any given moment, have the potential to limit the further development of its power to act, and by consequence, its actual existence. This limit determines the margin of variation or proportion of the expression of the given individual's power to act, which varies from a minimum, below which it would cease to exist (intensity = 0), to a maximum, which would only be limited by the extent to which its power to act is further integrated at any given moment in more composite relations. It is these kinds of relations that are expressive of the affective life of an individual.

Any resolution of this apparent contradiction requires an understanding of Spinoza's theory of relations, which deals with the relation between the singular modal essence of an individual human being and that human being's finite modal existence. In Macherey's introduction to the third part of the *Ethics*, when he speaks of 'the *conatus*' of a finite mode as constituting its 'actual essence' (E IIIP7), he writes that 'the power [*puissance*] . . . of the *conatus* of each thing is . . . distributed between a minimum and a maximum, the first corresponding to a pole of extreme passivity, the second to a pole of extreme activity' (Macherey 1995: 24). When he writes 'between a minimum and a maximum', he endorses an interpretation of singular modal essence which admits a certain margin of variation. In *Expressionism in Philosophy*, Deleuze writes that, with Spinoza, 'the relation that characterises an existing mode as a whole is endowed with a kind of elasticity' (Deleuze 1992: 222) and that singular modal 'essences or degrees of power always correspond to a limit (a maximum or minimum)' (1992: 204). Exactly what each of these interpreters understands by a maximum and a minimum, and exactly how they reconcile the singular modal essence and the finite modal existence of an individual with the concept of variation or proportion, is the focus of this chapter. The first part of the chapter provides an account of Deleuze's approach to this problem that clarifies the terms of the debate. The next part presents an account of Macherey's approach to the problem that establishes how their respective approaches differ. The final part identifies the key distinction that allows the explication of each of their respective solutions to the problem. The aim of the chapter is to establish that Deleuze's solution provides a more interesting schema by means of which Spinoza's *Ethics* can be used to model the relations that individuals have with things in the world.

Capacity to be Affected

The interpretation of modal essence, *conatus*, and the power to act that Deleuze offers us complicates that of Ramond. In *Expressionism in Philosophy*, Deleuze invites us not to 'confuse' the essence of a mode and the 'relation in which it expresses itself' (Deleuze 1992: 183). For Deleuze, when Spinoza defines modes as modifications of the attributes of substance, he refers only to singular modal essences, not to the existence of finite modes, which is determined solely by the effects of existing modes on each other. Singular modal essences, as modifications of an attribute, are characterised by Deleuze as being composed of 'intensive parts'. Each mode is composed, therefore, of both a singular modal essence, or intensive parts, within an attribute, and, corresponding to this, a finite existing mode. Deleuze understands intensive parts to be 'parts of power [*puissance*]', that is, 'intrinsic or intensive parts, true degrees', which are distinguished from one another as different 'intensities' or 'degrees of power' (1992: 173).

The notion of 'intensity' appears in the *Ethics* (E IVP9–12), but the expressions 'intensive parts' to designate modal essence, and 'degrees of power [*puissance*]', are never used by Spinoza. In his notes to the English translation of *Spinoza et le problème de l'Expression*, Deleuze writes that:

> It is quite true that one doesn't, strictly speaking, find intensity in Spinoza. But *potentia* and *vis* cannot be understood in terms of extension. And *potentia*, being essentially variable, showing increase and diminution, having degrees in relation to finite modes, is an intensity. If Spinoza doesn't use this word, current up to the time of Descartes, I imagine this is because he doesn't want to appear to be returning to a Precartesian physics. Leibniz is less concerned by such worries. And does one not find in Spinoza the expression '*pars potentia divinae*'? (Deleuze 1992: 417–18)

A thorough discussion of the problem of 'intensity' in *Spinoza et le problème de l'Expression* is inseparable from the distinction of qualitative and quantitative which serves to explicate the relation (*rapport*) of substance to its attributes and to its modes. Deleuze draws upon the work of Duns Scotus to develop a clear distinction between the measure of the intensity of a quality, which Deleuze characterises as an intensive quantity, and its quantitative extension, which Deleuze characterises as a corporeal or extensive quantity. There are therefore degrees or intensities of a quality, referred to in general as 'intensities' or 'degrees of power', which are just as different from quality as from extensive quantity. To support Deleuze's use of the notion of 'degrees of power', we can look to the *Ethics* itself, where Spinoza speaks of 'degrees'. In *Ethics* IIP13S, he writes that 'the things we have shown so far are completely general and do not pertain more to man than to other individuals, all of which, though in different degrees, are nevertheless animate [*omina . . . diversis gradibus animata . . . sunt*]'. Deleuze therefore employs the expression 'degrees of power' to define 'modal essence' with a certain legitimacy.

Deleuze declares that, for Spinoza, 'modal essences are [. . .] parts of an infinite series' (Deleuze 1992: 198). He argues that the essences of finite modes

> do not form a hierarchical system in which the less powerful depend on the more powerful, but an actually infinite collection, a system of mutual implications, in which each essence conforms with all of the others, and in which all essences are involved in the production of each. (1992: 184)

Therefore the degrees of power or essences of modes within any attribute are mutually determined by one another. Corresponding to any determined degree of power is an existing mode. Existing modes can be seen as extensive parts, which are external to one another, and which act on one another from the outside. An existing finite mode comes to exist by virtue of an external cause; its cause is another existing finite mode, whose own cause is another existing mode, and so on ad infinitum. The component parts of an existing mode are external to the mode's essence but these extensive parts exist directly in relation to the mode's intensive parts, that is, to its essence or degree of power. The relations between the extensive parts which constitute the existence of a mode or composite body, and which correspond to that mode's essence, are determined by 'purely mechanical laws' (Deleuze 1992: 209). Such a mode comes to exist when a number of extensive parts enter into a 'given' relation which corresponds to a given modal essence, when its parts 'actually belong to it [. . .] *in a certain relation of movement and rest*' (Deleuze 1992: 208).[2] Deleuze refers to this relation as 'a characteristic relation' (1992: 209). A mode continues to exist as long as the ratio or proportion of relations and relative movement of the extensive parts is maintained. An existing mode or composite body is thus open to continual alteration of its extensive parts, but it will continue to exist as long as the same ratio or proportion of extensive parts subsists in the whole (E IIL4–7). Therefore, a finite existing mode only exists insofar as a number of extensive parts are temporarily related to that essence. It is in this way that it is determined 'as having duration, as having a relation with a certain extrinsically distinct time and place' (Deleuze 1992: 213; see E IIP8C).

The duration of a mode's finite existence rests not only with the preservation of the proportion or ratio of its extensive parts, but also by the spatial and dynamic characteristics of the extensive parts belonging to it at any moment, 'and' by it being further integrated as a component part in more composite relations. The finite existence of a composite body is thus determined by its incorporation in an increasingly composite assemblage of relations, with only its characteristic relation enduring while the extensive parts of which it is constituted come and go. This is the concept of the individual that takes into account those relations with other bodies that lead to an increase in its power. Such an individual, as a whole, incorporates other individuals or bodies, and is itself incorporated as a part into more composite, although not necessarily more complex, individuals or bodies (Matheron 1969: 58).[3]

The spatial and dynamic characteristics determinative of the finite existence of a mode correspond to a certain 'capacity' of the mode's extensive parts 'to be affected' by other extensive parts. Such a determination by an external body is what Spinoza understands to be an 'affection' of the human body, that is, the way in which the human body is affected by another body. Deleuze argues that if an individual is able to undergo, or withstand, such affections without them changing the proportion or ratio of its extensive parts, that is, without the composite assemblage of relations of which it is composed being destroyed or decomposed by them, then this constitutes its 'capacity to be affected'. Insofar as a body is an individual human being, whose structure is constituted by the composition of an assemblage of relations, this assemblage of relations corresponds to that individual's capacity to be affected. Furthermore, Deleuze argues that an affection in Spinozist terminology may be passive or active, depending on whether the affection is determined from without by an external body, or whether it is determined by the mode's own degree of power.

Deleuze argues that to the extent that an individual's affections are passive, it is said to suffer, or undergo things, and its degree of power is expressed by a 'power of suffering' (1992: 222). To the extent that an individual's affections are active, it is said to act, and its degree of power is expressed by a power of acting. Deleuze writes that the 'capacity to be affected remains constant, whatever the proportion of active and passive affections' (1992: 222). However, within this fixed capacity to be affected, the proportion of active and passive affections is open to variation. The production of active affections will have as a result a corresponding reduction of passive affections, and, reciprocally, the continuation of passive affections will inhibit proportionally the power to act. Thus, for Deleuze, the power to act of an essence is open to variation.

When Deleuze first discusses the *conatus* of a mode he seems to argue that it remains fixed for any given mode. He writes that the *conatus* of an existing mode is 'the effort to preserve the relation [*rapport*] of movement and rest that defines it, that is, to maintain constantly renewed parts in the relation [*rapport*] that defines its existence' (Deleuze 1992: 230). However, rather than stating the strict equality between *conatus* and modal essence, Deleuze argues that a mode comes to exist when its extensive parts are extrinsically determined to enter into the relation that characterises the modal essence, and he emphasises that 'then, and only then, is its essence itself determined as a *conatus*'. 'A *conatus* is indeed a mode's essence (or degree of power) *once the mode has begun to exist.*' From this, Deleuze argues that *conatus* is the 'affirmation of essence in a mode's existence' (1992: 230). Both of these phrases seem to indicate that there is a relation different to a relation of strict equality between *conatus* and essence. But before attempting to determine what Deleuze means by this notion of affirmation, we will turn to Macherey's interpretation of modal essence, *conatus*, and power to act, and his determination of the relations between them, which contrasts markedly with Deleuze's account just developed.

Uninterrupted Affective Flux

Macherey bases his argument on the foundation that the *conatus* of a finite mode constitutes its 'actual essence' (E IIIP7). For Macherey, neither the *conatus* nor the power to act of a finite mode is variable. How then does Macherey incorporate a concept of variation into his interpretation of Spinoza? Macherey already accounts for the point of view held by Ramond, that the power to act of a singular thing is fixed, when he says that 'we are here very close to the wording/exposition of a problem of logic, which concerns essences insofar as they are fixed once and for all by definitions which they evidently cannot contravene' (Macherey 1995: 74). This problem of logic is resolved by what Macherey considers to be 'an economic perspective on the system of the affective life [*la vie affective*] as a whole' (1995: 24). According to Macherey, the affective life of a mode is characterised by the power (*puissance*) of its *conatus*, which is deployed between the poles of extreme passivity and activity. This can be characterised in part by Spinoza's theory of the passions.

Macherey begins his analysis of the affective life by asking 'What is an affect?' It is in Definition 3 of *Ethics* III that Spinoza defines what he means by affect. Macherey explains this in the following way: 'it is the *idea* of an affection of the body which corresponds to an augmentation or diminution of its power to act: this variation is in relation to the fact that the body is affected, sometimes by itself', resulting in the idea of an augmentation of its power to act, and 'sometimes by an external body' (Macherey 1995: 355), resulting in the idea of a diminution of its power to act. In the General Definition of the Affects, Spinoza says that an affect is a confused idea by which the mind affirms of its body, or some part of it, a greater or lesser force of existing than before, explaining that:

> when I say a greater or lesser force of existing than before, I do not understand that the Mind compares its Body's present constitution with a past constitution, but that the idea which constitutes the form of the affect affirms of the body something which really involves more or less of reality than before. (E IIIDef.Aff.)

Macherey argues that this *idea* expresses only 'a momentary state of our body ... in rupture with [its] preceding state' (Macherey 1995: 355), in the sense of an augmentation or diminution, which marks a development or a restriction of its power to act. As Spinoza says, all affectivity is based on the foundation of 'joy' and 'sadness' which, according to Macherey, expresses the 'transformations' (*mutationes*) associated with the fact that the soul (or mind) is, without end, exposed to 'passing sometimes to a greater sometimes to a lesser perfection'. These transformations, or 'transitions' (*transitio*), are experienced as 'passions', at the heart of which the 'soul', Macherey writes, 'is completely subject to the mechanisms of the imagination' (1995: 121). The fact of passing to a greater perfection leaves the soul 'euphoric' (1995: 333). However, Macherey emphasises that even though this 'joy' experienced by the soul is perfectly real, the base upon which it rests remains imaginary.

The idea of a mode's perfection of which Spinoza speaks is a measure of that proportion of the mode's power to act which is expressed actively by this mode at any precise moment. In accord with the concept of a mode's fixed essence and power to act, Macherey suggests that whatever the margin of variation within which the power (*puissance*) of the soul is expressed, it 'remains in all cases, and by definition, the same' (1995: 125), and the soul 'maintains this conformity to its nature without which it would simply cease to be' (1995: 123). Therefore, to pass to a greater perfection is to express 'actively' a greater proportion of a mode's fixed power to act.

Passing to a lesser perfection is neither to be deprived of a greater perfection, in the sense of a lack, nor is it 'a pause in the pursuit of the movement which effects the fundamental impetus of the *conatus*' (Macherey 1995: 124). Sadness is the inverse of joy, but it is not the absence of joy. As Macherey writes, sadness is 'a contraction of the power to think of the soul, which deprives it momentarily, but which would be unthinkable without the persistence of this power, of which it continues to give a paradoxical expression' (1995: 125). Sadness therefore is an integral part of the movement of the *conatus*. What Macherey wants to make understandable by reasoning in this way is that, whatever the orientation of the variation which affects the soul, whether joy or sadness, the soul continues in all cases to be animated by the pressure of the *conatus*, and the 'transitions' that the soul does not cease to experience in no way alter the constancy of the *conatus*. It is, writes Macherey, 'constitutionally inalterable, and persists imperturbably across these series of transitory states' (1995: 125). In postulating that the soul can be confronted at any moment by the alternative between two contrary orientations, Macherey maintains that there is no room to think that the soul 'would have more control of itself when it is occupied by those sentiments which elate it than when it is occupied by those sentiments which depress it' (1995: 121). The character of the 'transitions' therefore leaves the affections 'fundamentally ambiguous' (1995: 125). As none of these states of the soul have in themselves a guarantee of stability, the affective life remains in a 'state of permanent instability' (1995: 125). This state, Macherey suggests, prevails in what he refers to as the 'uninterrupted affective flux [*le flux affectif ininterrompu*], which varies continuously between the two extreme poles of a maximum and a minimum' (1995: 121).

Macherey concludes that the states of activity and passivity of the soul are not absolute states, and as such are not radically exclusive one of the other. They are, rather, measured against one another at the interior of a gradual series of states described as the 'uninterrupted affective flux', which tends to realise all intermediate states between the two extremes. The variation of the active expression of a mode's power to act is therefore inversely related to the proportion of a mode's whole power to act that is occupied by the 'uninterrupted affective flux'.

Passive Affections

While Macherey analyses affectivity from the point of view of the imagination, in relation to the attribute of thought or the soul, Deleuze's explication remains with the point of view of the attribute of extension, in relation to the body's power to act.[4] Macherey considers the effect of individual affects, whether joy or sadness, to remain fundamentally ambiguous, absorbed within the 'uninterrupted affective flux', which functions as a hindrance or limit to the active expression of a mode's fixed power to act.[5]

Deleuze's interpretation of Spinoza's theory of the affections, however, introduces another level of variation. Even though Deleuze considers the capacity to be affected of a 'given' mode to be fixed, he maintains that it does not remain fixed at all times and from all points of view. 'Spinoza suggests', Deleuze maintains, 'that the relation that characterises an existing mode as a whole is endowed with a kind of elasticity' (1992: 222). The individuation of a mode changes on leaving behind childhood, or on entering old age, and also after the permanent effects of illness.[6] Such changes as growth, ageing, and illness may be understood as though a mode's capacity to be affected, which corresponds both to the composite assemblage of relations of which it is composed and to the more composite relations in which it is further integrated, enjoys 'a margin, a limit within which' these global relations 'take form and are deformed' (Deleuze 1992: 223). What Deleuze understands by this is that, even though quantitatively there has been a change in the way that the singular modal essence of an individual is actualised in a composite relation, qualitatively, insofar as the singular modal essence belonging to this individual is a modification of an attribute, it remains unchanged. So although the individual's singular essence remains unchanged, there is a change in the individual's capacity to be affected, which corresponds to a change in the extent to which it is further integrated as a component relation, in larger, more composite relations.

This distinction between the fixed, singular, modal essence of an individual and its changing capacity to be affected can be characterised by differentiating Deleuze's understanding of the effect of the power of suffering from that of Macherey. Deleuze argues that the power of suffering and the power of acting of a finite existing mode can only be considered as two distinct principles, which are inversely proportional to one another within a 'given', or fixed, capacity to be affected, insofar as the affections are considered 'abstractly, without concretely considering the essence of the affected mode' (1992: 223). Macherey's theory of the affective life as a state of 'uninterrupted affective flux' remains abstract in this way, in so far as it solely considers the affections of a finite existing mode abstracted from the nature of the relations in which it is involved. Although Deleuze maintains that this abstracted theory is a primary thesis of Spinoza, he suggests that 'this thesis, if physically true, is not metaphysically true' (1992: 224). What Deleuze means is that even though it might be the case that all finite existing modes in general are involved in relations in which they can be considered to act and in those in which they suffer, when considered from the point of view of any particular finite modal existence, and how they might understand the relations in which they are involved, they might well know that they suffer, but they cannot be said to understand the relations that they suffer nor can they be said to be constituted by those relations. Deleuze maintains that, for Spinoza, the power of suffering expresses nothing positive or real. That proportion of a fixed capacity to be affected that suffers relations with objects whose natures disagree, or are not compatible, with our own is inhibited from being expressed by virtue of those relations. The idea that this mode might have of this object is partial, and therefore imaginary. Deleuze characterises such a relation as follows: 'in every passive affection there is something imaginary which inhibits it from being real' (1992: 224), that is, which inhibits it, the passive affection, from being real for the finite existing mode itself. Macherey is in agreement with Deleuze on this point. However, despite this point of convergence in their interpretations of Spinoza's physics, Deleuze further develops Spinoza's theory of the affections differently to that presented so far by Macherey in order to bring out this distinction between the fact of suffering in general and the perspective of a particular finite existing mode and its understanding of the relations in which it is involved.

The difference between their respective accounts of Spinoza's theory of the affections hinges on the difference in how they each conceive of passive affections to function as limits. The way that 'the affections', in a state of 'uninterrupted affective flux', hinder or 'limit' the expression of a mode's power to act is very different from the way that passive affections 'limit' the expression of active affections. For Macherey, the term 'limit' functions to explain the impact of 'the affections' on the full expression of a finite existing mode's power to act, the uninterrupted affective flux thereby limiting the active expression of the mode's power to act to only a proportion of its range, which varies between a maximum and a minimum, that is, to less than maximum. However, for Deleuze, the term 'limit' defines a margin or threshold beyond which a mode's capacity to be affected ceases to be animated by active affections and therefore ceases to be expressed altogether, that is to say, beyond which a finite existing mode ceases to exist. This distinction between the ways in which the concept of 'limit' is used is key to identifying the difference between their respective accounts of modal existence. In order to further clarify this distinction, I will graphically map its implications for each of their accounts.

Ramond criticises Deleuze for using the term 'limit' in the singular ('*a* limit', writes Deleuze, '*a* margin') 'since what is actually designated are two limits ("a maximum *and* a minimum")', Ramond argues, 'or two margins, "between which", Deleuze should have written, and not ... "in which" the bodies "take form and are deformed"' (Ramond 1995: 226). This criticism underlies the difference between the two concepts of 'limit' that are used by Macherey and Deleuze respectively. Macherey considers there to be two limits, a maximum and a minimum. According to him, the active expression of a mode's power to act is hindered by the 'uninterrupted affective flux' to vary within these finite limits, whereas Deleuze uses the term 'limit' in the singular to define a point beyond which an existing finite mode ceases to exist.

One might ask how Deleuze can speak of a limit in the singular when he also speaks of a maximum and a minimum in what appears to be a similar way to Macherey. Deleuze here draws upon the distinction mentioned earlier between what might be 'physically true' in general about finite modal existence – that is, the fact that all finite existing modes are involved in relations in which they can be considered to act and suffer – and what he claims is 'metaphysically true' (Deleuze 1992: 224) of particular finite existing modes and how they might understand the relations in which they are involved. They might well know that they suffer, but they cannot be said to understand the relations that they suffer, nor can they be said to be constituted by those relations. Deleuze distinguishes between these two different perspectives by referring to the former as the 'physical view' and the latter as the 'ethical view' (1992: 225). Deleuze argues that it is only in what he refers to as the 'physical view' that the capacity to be affected remains fixed for a given essence, and can therefore be represented by a fixed range of variation between a maximum and a minimum. Macherey's point of view of modal existence fits this description of the physical view (see Figure 6.1). In the diagram, the limit to the active expression of a mode's power to act is represented by the inverted triangle, which splits the fixed power to act of the

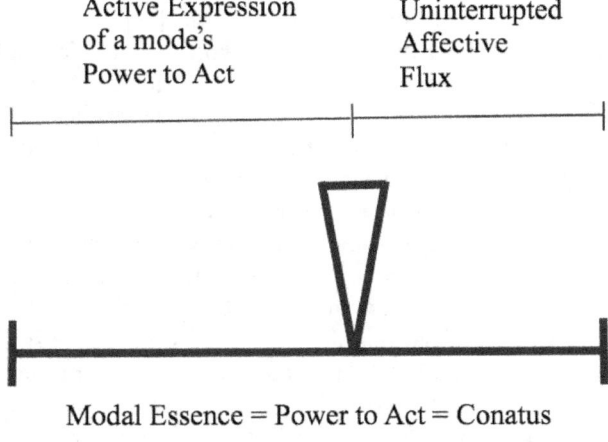

Figure 6.1 The physical view

modal essence into two parts, the other being the uninterrupted affective flux.

However, Deleuze introduces another point of view which he calls the 'ethical view', in which a mode's capacity to be affected 'is fixed only within general limits' (1992: 225). Before determining the difference between these 'general limits' and 'the limit' of which he speaks in the singular, it is necessary to explicate Deleuze's 'ethical view'. What distinguishes the ethical view from the physical view is a further qualification of the nature of the distinction between passive and active affections. In the ethical view, passive affections function solely as a limit to the existence of a finite mode. They no longer express modal essence, as all passive affections continue to do on the physical view with Macherey, but actually limit its expression. What differs is the definition of a mode's capacity to be affected, which was understood on Macherey's physical view to be the combination of a mode's power to suffer and its power to act. According to the ethical view, the power to suffer is no longer considered to be expressive of a mode's capacity to be affected. A mode's capacity to be affected is rather expressed solely by its power to act. Deleuze describes the effect of the passive affections on a mode's capacity to be affected in the following manner: 'While exercised by passive affections, it is reduced to a minimum; we then remain imperfect and impotent, cut off, in a way, from our essence or our degree of power, cut off from what we can do' (1992: 225). The passive affections, which, according to the physical view, contribute to a mode's power of suffering, now only reduce or limit a mode's power to act to its '*lowest degree*' (Deleuze 1992: 224). Insofar as active affections express a mode's power to act, they are the only affections which exercise this new ethical concept of a mode's capacity to be affected. 'The power of action is, on its own,' Deleuze argues, 'the same as the capacity to be affected as a whole', and this newly defined power to act, 'by itself, expresses essence' (1992: 225). Therefore, according to the ethical view, an existing mode's essence is expressed by its power to act, and its power to act is the same as its capacity to be affected.

To see how *conatus* functions according to the ethical view we must take up the previous reference we made to Deleuze's understanding of the relation between *conatus* and essence. As we have seen, Macherey holds that there is a strict equality between these two aspects of a finite mode. He would therefore not disagree with Deleuze's statement that the '*conatus* is [. . .] always identical to the power of acting itself' (Deleuze 1992: 231), since Macherey also considers the relation of equality to hold between the power to act and modal essence. The difference between these respective interpretations of *conatus* comes out when we see how Deleuze interprets *conatus* in the ethical view. When Deleuze states that the *conatus* is the 'affirmation of essence in a mode's existence' (1992: 230), what he understands by this is that *conatus* is the affirmation of the expression of a singular modal essence in the corresponding finite existing mode only insofar as the finite mode actually exists, that is, only 'once it has begun to exist' (1992: 230). Now, it has been determined that a finite mode only exists for Deleuze insofar as it has a capacity to be affected, which is determined by the extent to which the composite assemblage of relations, of which it is composed, is further integrated in more composite relations. A mode's *conatus* is, therefore, the affirmation of its

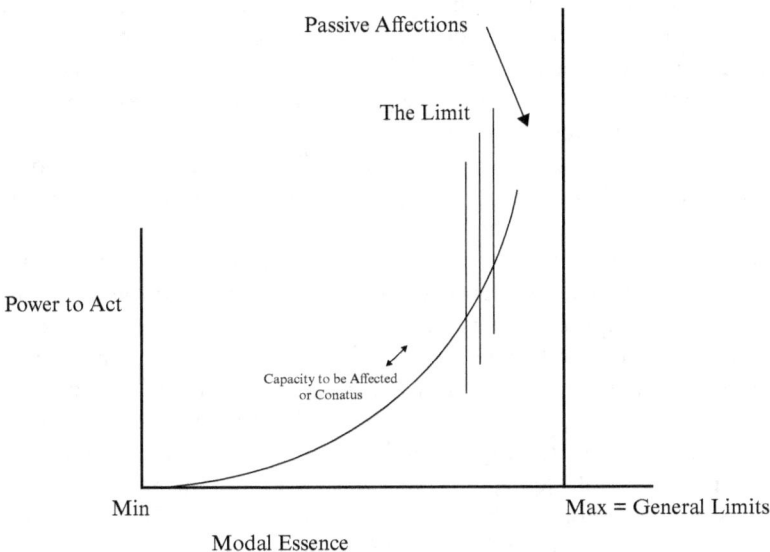

Figure 6.2 The ethical view

capacity to be affected, that is, of the extent to which it is further integrated as a component relation in more composite relations. Insofar as a mode's *conatus* is the 'effort to maintain the body's ability to be affected in a great number of ways' (1992: 230), which, for Deleuze, is the effort to maintain its capacity to be affected, it is the effort to maintain the extent to which it is further integrated in a great number of ways. The greater the extent to which the relation characteristic of a mode is further integrated, the more composite are the relations in which it is implicated; the greater is its capacity to be affected, and, consequently, the greater is its power to act. The *conatus* of a mode thus affirms the expression of its capacity to be affected as its power to act. Deleuze is therefore able to conclude that 'the variations of *conatus* as it is determined by this or that [active] affection are the dynamic variations of our power of action' (1992: 231). The *conatus* therefore affirms the expression of the dynamic variations of a mode's power to act. Figure 6.2 charts the potential for an exponential increase in a mode's capacity to be affected, along the curved line, which corresponds to an increase in its power to act, along the y-axis. While this occurs within fixed general limits of a mode's essence, indicated on the x-axis between minimum and maximum, the more composite relations that the mode has, the greater its capacity to be affected and therefore the greater its power to act. These relations are limited at any particular time by 'the limit', which simply represents the extent to which the mode is involved in such relations at any particular time.

Deleuze argues that finite modes are open to three different types of expressive changes. To begin with there are mechanical changes in the affections that a finite mode experiences. These changes are synonymous with the changes envisaged by Ramond in his argument about the precisely determined relation of movement and rest of a finite mode, a relation that remains the same as before despite the variety of movements and displacements undergone by its parts. All three interpreters agree with this interpretation of Spinoza's concept of modal essence as fixed. There are also dynamic changes in a finite mode's capacity to be affected. Here Deleuze's interpretation differs from that of Ramond and Macherey. For Macherey, dynamic changes are embodied by the varying extent to which the 'uninterrupted affective flux' inhibits or limits the active expression of a mode's power to act to between the range of a maximum and a minimum. All of a mode's power to act is expressed; however, according to the uninterrupted affective flux, it is expressed both actively and passively. Deleuze's concept of a finite mode's capacity to be affected, as we have seen, always already involves a concept of variable power to act as the expression of a mode's active power of existence. This arrangement is complicated further in the ethical view, when Deleuze directly equates the capacity to be affected with the mode's power to act. Passive affections, for Deleuze, now function as a limit of the expression of active affections, and therefore of the existence of the finite mode itself. This limit functions within the range of the given finite mode's fixed power of existence, or essence – that is to say quantitative essence, as intensity or degree of power – and therefore within a maximum and a minimum. Therefore, for Deleuze, the three aspects of a mode, its capacity to be affected which is expressed by its *conatus* as its power to act, are all together correspondingly open to variation within this range, and their expression is limited by the passive affections that they are together subject to.

However, Deleuze argues that a maximum and a minimum, or the range of variation, established by our power of existence function only as 'general limits', because finite modes are also open to a third type of change, what he calls 'metaphysical' changes. These changes are those which Deleuze refers to when he writes that modes 'are endowed with a kind of elasticity' (1992: 222). In the physical view, as far as Deleuze is concerned, the maximum and minimum of the range of variation determined by a mode's degree of power function only as 'general limits'. In the ethical view, Deleuze emphasises instead that 'while a mode exists, its very essence is open to variation, according to affections that belong to it at a given moment' (1992: 226). As has been noted, Deleuze argues that a singular modal essence is only fixed qualitatively, insofar as it corresponds to the qualitative modification of an attribute in the form of a complicated intensive quantity. It nevertheless remains open to variation under certain conditions quantitatively, insofar as its singular modal essence is further integrated in a more composite relation. The maximum and minimum are determined as 'general limits' because Deleuze uses the term 'limit' in the singular to indicate that a finite mode is not so much limited between a maximum and minimum, as it is by the passive affections that it experiences in its interactions with other more composite bodies. These, at any given moment, have the potential to limit its further integration, and, therefore, the further deployment of its power to act, and by consequence, its actual existence. Passive affections, for Deleuze, therefore not only function as a limit of the expression of a mode's active affections, but also of the existence of the finite mode itself. This limit determines the margin of variation of the expression of the given finite mode's power to act, which varies from a minimum, below which it would cease to exist (intensity = 0), to a maximum, which would be the extent to which its power to act is further integrated at any given moment in more composite relations.

Composite Relations and the Power to Act

Macherey and Deleuze therefore come to different conclusions concerning the resolution of the apparent contradiction between the double point of view of the singular essences of finite existing modes, according to which sometimes the power to act of a finite existing mode seems to be fixed to a precisely determined degree, whereas sometimes it seems to admit a certain degree of variation. According to Macherey, this contradiction is dissipated as soon as the fact that the two theses in question are not situated on the same plane is taken into account. He considers the concept of the margin of variation of a mode's power to act to be a problem of logic, which is resolved by taking an economic perspective on the system of the affective life of a mode as a whole. Macherey argues that a mode's affective life limits the active expression of its power to act, the associated ideas of the augmentation and diminution of which are experienced only at the level of the imagination, while its essence, *conatus*, and power to act remain fixed once and for all.

Deleuze considers the problematic of the apparent contradiction to be determined by two different perspectives within the *Ethics*: a physical view and an ethical view. Deleuze agrees with Macherey with regard to the perspective of the physical view. But the differences in their respective interpretations are determined by the different role given to the passive affections. For Macherey, they remain an integral part of the mode's existence, being expressed by the *conatus* of a mode although hindering its ability to express fully, or more perfectly, its fixed power to act. Deleuze's ethical view serves to resolve the apparent contradiction differently. One of the fundamental aspects of the distinction between the physical view and the ethical view is a change in perspective of the relation between passive affections and active affections. In the ethical view, only active affections function integrally as constitutive of finite modal existence. The active affections are solely determinative of the mode's capacity to be affected, which is affirmed by its *conatus* as the expression of its power to act. Passive affections function solely to limit the expression of the mode's power to act, and therefore of its existence.

By affirming the variability of a mode's capacity to be affected, as expressed in its power to act, while maintaining the concept of its fixed singular modal essence, Deleuze's ethical view resolves the apparent contradiction in Spinoza's *Ethics*; in fact, according to the ethical view there is no contradiction. What this means is that, rather than simply providing a schema by means of which the power to act of a finite existing mode, or individual, can be abstractly determined at any moment in time, which is afforded by the model of the physical view and those who subscribe to it exclusively, Deleuze extends this schema to include an account of the composite relations in which a finite existing mode, or individual, is specifically involved that result in an increase in its power to act. What Deleuze calls the 'ethical view' thus represents a schema by means of which Spinoza's *Ethics* can be used to model relations that individuals have with things in the world that is otherwise only indirectly, and insufficiently, accounted for.

Notes

1. For a more extensive treatment of Ramond's reading of Spinoza on this issue, see Duffy (2006: ch. 2).
2. The foundation of Spinoza's physics appears in Proposition 13 of Book II of the *Ethics*, where he defines the essence of a body by a *certa ratio motus et quietis*: a certain or precise relation of movement and rest. The 'precise relation' of 'the essence of a singular thing' is indifferent to certain types of variation. Nutrition, growth, and all forms of movement and displacement in space are able to be produced without altering the individual essence. 'By this, then,' writes Spinoza, 'we see how a composite individual can be affected in many ways, and still preserve its nature' (E IIL7S).
3. In the letters to Boyle (Ep. 6, 11, 13, and 16), Spinoza expresses his doubts concerning the mechanist explanation of phenomena, which supposes the hierarchy from simple to complex, or the idea of a hierarchy of complexity based upon a hierarchy of composition.
4. 'By affect I understand affections of the Body by which the Body's power of acting is increased or diminished, aided or restrained, and at the same time, the ideas of these affections' (E IIID3).
5. Macherey bases his conception of the functioning of the 'uninterrupted affective flux' on E IIIP17, where Spinoza introduces the theme of affective ambivalence, *fluctuatio animi* (Macherey 1995: 163).
6. '[W]hat shall we say of infants? A man of advanced years believes their nature to be so different from his own that he could not be persuaded that he was ever an infant, if he did not make this conjecture concerning himself from [NS: the example of] others' (E IVP39S). See Deleuze (1992: 222).

7

Spinoza, Heterarchical Ontology, and Affective Architecture

Gökhan Kodalak

Spinoza's philosophy offers a radical conception of life that traverses nature and culture, humans and non-humans, built and natural environments, promising far-reaching implications for architecture. The relationship between Spinoza and architecture, however, has been nothing but a huge missed encounter. Despite the promise of recently emerging scholarship,[1] not a single monograph has been published on this relationship for almost three and a half centuries. We may ascribe this missed opportunity to architects' incidental neglect, or perhaps even deliberate disregard, of Spinoza's philosophy for denying them the privilege of situating themselves as hegemonic shapers of the built environment; or to Spinoza's multi-layered language and convoluted conceptual framework that resist easy translation; or even to Spinoza's controversial reception as a heretical figure and a subterranean philosopher throughout modernity. Regardless of what the motives may be, what is overlooked here is that there is much more to the relationship between Spinoza and architecture than meets the eye. My hypothesis is that there is a latent architectural treatise underlying Spinoza's entire oeuvre, including his private letters. But much like his unfulfilled promise of a treatise on physics (Ep. 83 to Tschirnhaus, 1676), or his unfinished political treatise, Spinoza's treatise on architecture is not to be found as a ready-made manuscript. Rather, its unravelling requires the discovery of discontinuous spatial hints and subtle architectural connotations buried deep between the lines in his philosophical archive. An introductory exploration of the promising relationship between Spinoza and architecture constitutes the subject matter of this chapter, an exploration that interweaves these fragmentary hints with a confluent architectural lexicon, orienting us towards a peculiar journey with the potential to redefine all the familiar terms we take for granted at the intersection of philosophy and architecture.

Onto-epistemology

At the outset of the *Ethics*, Spinoza introduces a new way of seeing reality, a new way of conceiving the cosmos, a new way of grasping life (E ID1–8). This novel vision is presented with the meticulous conceptual elaboration of an onto-epistemological triad consisting of substance, attributes, and modalities.

Substance is the singular continuum through which all individual differences unfold. As part of his overall strategy, Spinoza appropriates common terms and calls substance *Deus sive Natura*, that is, God or nature (E IVPref.; KV App.II). Here, *sive* ('or') expresses a seemingly innocent but subtly radical gesture that redefines both terms together. For Spinoza's substance is neither a sovereign God creating the world from a transcendent dimension, nor a mechanical nature merely consisting of the sum of all extensive individuals. Rather, substance is the fullness of reality, the immanence of life potentiating infinite capabilities unfolding in manifold planes of heterogeneous existence (E IP14C1; CM II.6). Attributes are expressions of substance. The cosmos expresses itself via an infinity of attributes, two of which are known to us as extension and thought (E IP11; E IIP1–2). When extension is expressed as a specific body, or thought as a particular mind, new modalities of life emerge. These modalities, lastly, are durational modifications of substance (E IP25C).[2] Finite modalities are what we are, what buildings, tornadoes, black holes happen to be. A singular substantial continuum that expresses itself via an infinity of attributes which are enveloped in a multiplicity of finite modifications: this is Spinoza's onto-epistemological triad.

What holds this cosmogenetic triad together is Spinoza's concept of immanence. In the history of medieval thought, the arrival of immanence was already anticipated in Ibn Arabi's wahdat al-wujud (1980: 26) and Duns Scotus's univocity (1987: 82–96). Yet what Spinoza does is to push immanence to its conceptual limit, abolishing the conventional bifurcation of reality into creative and created

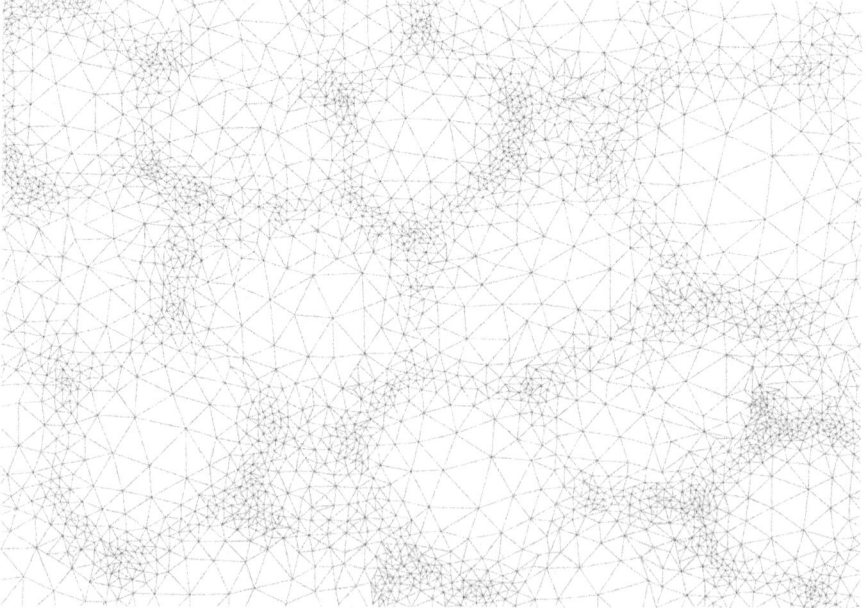

Figure 7.1 *Natura naturans*: substantial expression of life

dimensions, and ruling out any form of transcendent dimension filled by sovereign entities residing, deciding, or imposing immutable rules from above and beyond the immanent interactions of this cosmos. In Spinoza's philosophy, a consistent critique of transcendence always goes hand in hand with the perpetual refinement of immanence. He is well aware that transcendence is a notion with a tendency to infiltrate everywhere; it is not utilised solely by theological doctrines or absolute monarchies. Transcendence is an empty seat that is filled whenever a set of rules, regulations, value judgements, or power relations require absolute compliance. And it is this practical character of transcendence that has established long-standing alliances among architecture, philosophy, and theology, rendering the architect, in countless narratives throughout history, the transcendent archetype *par excellence*. With this enduring alliance, we have constructed an archetypal Platonic architect, who gives form and order to unformed chaotic matter according to ideal templates and transcendent proportions (Plato, *Timaeus* 29a–b [2008: 17]). We have constructed an archetypal Scholastic architect whose omnipotent agency creates matter *ex nihilo*, and who, by shaping this inert mass from beyond, crafts everything in harmony (Kostof 1977: 79). We have constructed an archetypal Renaissance architect whose genius imitates the absolute perfection of design (*disegno*) applied to the cosmos by 'the Divine Architect of Time and Nature' (Vasari 2008: 3–6). And we have constructed an archetypal Cartesian architect, who paves the way for modern successors by conceiving analytical plans and constructing carefully made devices with self-conscious authorship and rational will, distinct from and above the so-called blind functioning of the universe (Descartes 1985: 205). Eternal ideas, proportions, and forms; omnipotent form-givers and top-down designs; creative genius and supreme rational will. These are some of the transcendent gestures that have constituted the mythic aura of architecture, with which most architects cannot resist surrounding themselves even today.

Yet where others see various opportunities to fill the seat of transcendence and assign external creators and monarchs to life, Spinoza declares that the power to cause effects, and bring about existence, is immanent to those effects and existents; that is, life is immanent only to itself (E IP18; KV I.3–4). This means that the cosmos exists via a complex combination of immanent, efficient, and distributed causality, producing its effects on the same plane of its interactions. Each modality is a different modification of this cosmos, a unique composition of life's infinite capabilities. Nothing creates life from beyond, nothing shapes matter from above. There is no transcendent principle to the cosmos, no external cause to any creative process. Instead, each creative process consists of an implication or enfolding of life forces on the same plane of immanence, a complication or co-folding of these forces within modal meshworks, and an explication or unfolding of new compositions through the alembic of modal interactions. So, what will become of us architects, if we can no longer maintain the illusion of being transcendent creators of the built environment, if we can no longer impose form on so-called inert materiality with our omnipotent agency and top-down determination? Spinoza's immanent lens renders visible alternative paths: surfing

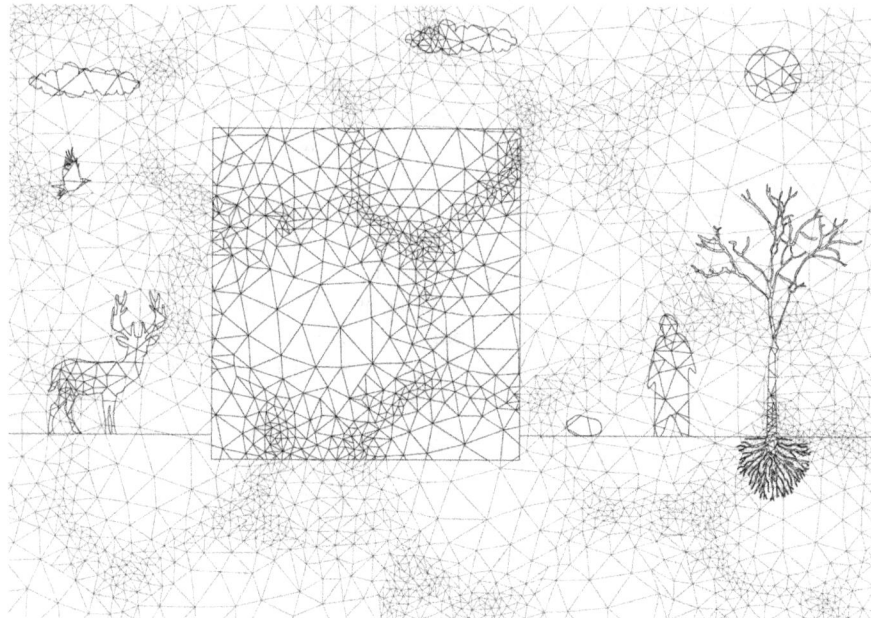

Figure 7.2 *Natura naturata*: modal expression of life

architectural capacities rather than commanding them, meeting spatial tendencies halfway rather than mastering them, negotiating morphogenetic potentials and augmenting architectural interactions rather than pre-determining and fixing them.

Taking immanence to its logical conclusion, Spinoza defines individual modalities no longer according to general kinds or universal essences, but via their affective capabilities and singular powers. Universals and general essences are a fallacy, Spinoza proclaims, originating in Platonic and Aristotelian traditions as 'they have set up general Ideas, with which, they think, particular things must agree' (KV I.6). Essentialist and universalist definitions look for the eternal underlying the accidental, or the universal beneath the singular, to retroactively impose a homogeneous set of innate constants on the unique lives of heterogeneous individuals. Spinoza objects that 'it is precisely the singular things, and they alone, that have a cause, and not the general, because they are nothing'. And hence, 'Peter must, as is necessary, conform to the Idea of Peter, and not to the Idea of Man' (KV I.6). This is Spinoza at his most provocative, telling us that we do not even need to conform to being human. For we are not defined by pre-determined generalisations such as gender, ethnicity, or even species, but by our singular potence (*potentia*), actualised in our everyday modifications. In other words, you are not defined by what you are (by your given membership of universal classifications), but by what you can do (by how you potentiate your entangled existence). Spinoza's approach suggests a novel way of defining architectural modalities as well, no longer as members of ready-made types, products of an overarching zeitgeist, or instances of identifiable styles, but according to unique interactions that they mobilise, singular modes of existence that they potentiate, specific performances that they cultivate.

In Spinoza's philosophy, ontology and epistemology are inextricably coupled. Spinoza finds Cartesian dualism unacceptable for presenting mind and matter as two distinct substances existing independently from each other (PPC IP8; E IIP7). Descartes uses this distinction to valorise mind as the only locus of creative activity while reducing material bodies to passive automata, rendering nature incapable of non-mechanical novelty and paving the way for modern anthropocentrism. Cartesian dualism replaces the conventional function of souls with the modern function of minds, heralding the subsequent repercussions of the Enlightenment's dark side, such as rendering mind and rationality the disciplinary protectorate of body and sensuality, and constructing hierarchical chains based on reason and progress for sexist, racist, and colonial forms of domination. Conversely, Spinoza forbids any form of supremacy that situates mind over body. His response to the mind–body problem falls neither on the rationalist side of subordinating body to mind, nor on the materialist side of reducing mind to body. From Spinoza's perspective, we do not have minds or bodies, but we are minds, we are bodies. That is, mind and body are 'one and the same thing, conceived now under the attribute of thought, now under the attribute of extension' (E IIIP2S). Mind and body are different expressions of the same psychosomatic event, confluent in all modalities without exception (E IIP13S). This means that architectural modalities also contract knowledge, process data, and exchange information with other modalities to the degree of their compositional complexity. Architectural buildings do not only express themselves through material formation, but also via immaterial information.

This brings us to the question of how modalities of existence are distributed in life. The scholastic model of the Great Chain of Being ranks entities according to a pre-defined hierarchical order (Lovejoy 2001). In this model, hierarchy means a vertical distribution of power (*arche*) according to an immutably sacred partition (*hieros*). God is commonly depicted at the top of this chain, progressing downward to angels, humans, animals, plants, and non-living things. This means that some entities have more being and reality, and are more valuable, than others, according to their ontological position on this hierarchical ladder. In Spinoza's lifetime, humanist and Enlightenment thinkers had already started modifying this hierarchical chain by placing humans at the top of the scale, yet keeping the descending ladder towards animals, plants, and non-living things intact. Accordingly, Descartes (1985: 131–41) declares that things are passive entities obeying only mechanical rules, animals are brute organic automata lacking reason, whereas humans, with their special cognitive traits, constitute an exceptional kingdom above all others. What is noteworthy about these hierarchical chains is their continuing prevalence in modern philosophy, science, and everyday imagination. In the early twentieth century, Martin Heidegger (1995: 176–8) insisted that 'the stone is worldless [*weltlos*], the animal is poor

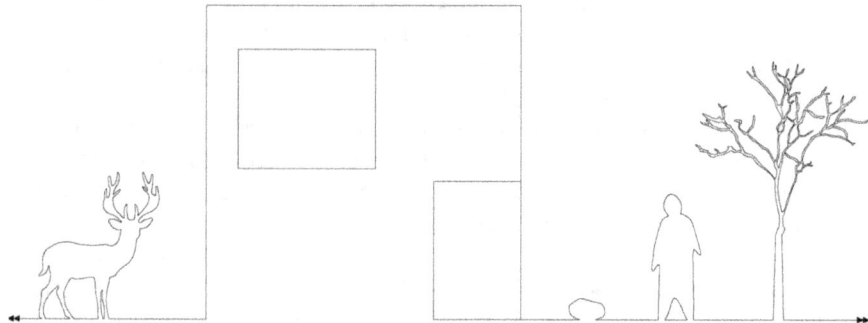

Figure 7.3 Heterarchical plane

in world [weltarm], man is world-forming [weltbildend]'. Even today, prominent scientists including Stephen Hawking (1998: 128–31, 142) believe in 'the Anthropic Principle', which maintains that the cosmos is fine-tuned for humans, so that we may witness and discover the universe's marvels. The fascination with hierarchical chains has been far-reaching. Jacques Derrida (1998: 573) warned architects that 'an always hierarchizing nostalgia' has haunted architecture from time immemorial, enticing the profession to ground itself on these vertical chains to the point of 'materializ[ing] the hierarchy in stone'.

Spinoza's ontology is vastly different, one that we may call a heterarchical plane, rather than a hierarchical chain. Here, heterarchy means that every modality equally exists on the same plane, but with heterogeneous compositions (*heteros*) and singular capacities (*arche*). In other words, ontological differences are not vertical but horizontal, in what Manuel DeLanda calls a 'flat ontology' (2002: 47). Each individual modality – whether human, animal, plant, or building – shares the same ontological reality with others. There are no ontological monarchs or castes. Nothing is privileged or elevated by essence. Nothing has more intrinsic being or value. Spinoza insists not only that 'God has no special kingdom over people' (TTP ch. 19/G III 229), but also that humans do not constitute a 'kingdom within a kingdom' (E IIIPref.). This is Spinoza's abolition of theocentrism and anthropocentrism at once. This is the abolition of ontological hierarchy as such. Instead, Spinoza's heterarchical plane is pan-affective in composition; it acknowledges the agency of each and every modality and affirms their equal share in existing, acting, affecting, and making a difference. Nothing is inert or passive: all modalities are 'animate albeit in different degrees' (E IIP13S). This means buildings are no longer at the bottom of a hierarchical ladder as non-organic entities; no longer devoid of agency, lacking activity and affective power; no longer worldless, dormant and static; no longer mere reflections of social conventions or their creators' intentions. Rather, buildings are world-forming. They do have an active vitality of their own. They do harbour unique capacities that affect the course of events. From Spinoza's heterarchical lens, architectural modalities are alive.

Modal Individuation

This brings us to the question of modal individuation. Although Spinoza could not elaborate his physics in full force, latent connections between his concepts of simplest bodies and motion-and-rest give us a clue about the individuation process he had in mind. According to Spinoza, an individual consists of the interaction of an infinity of simplest bodies (E IIA2″D). These simplest bodies, however, should not be mistaken for indivisible building blocks endowed with form and magnitude like atoms. Rather, they can be interpreted as formative forces underlying differential relations of motion-and-rest. Motion has a dual meaning here. The first appears to be drawn from Cartesian physics: motion as the transfer of one part of matter from one location to another. This is motion expressing the kinetic process of spatial displacement. The second is Spinoza's distinctive invention: motion-and-rest as an immediate infinite modality harbouring infinitesimal multiplicities that constitute differential relations through which individual modalities come to be and differ from each other (Ep. 64 to Schuller, 1675). This is motion expressing the dynamic process of individuation and modification. Spinoza accentuates this undertone as early as his *Short Treatise* (KV IIPref.):

> All and sundry particular things that are real, have become such through motion-and-rest . . .
> The differences among these result solely from the varying proportions of motion-and-rest . . .
> From such proportion of motion-and-rest comes also the existence of our body . . .
> This body of ours, however, had a different proportion of motion-and-rest when it was an unborn embryo; and in due course, when we are dead, it will have a different proportion again.

This means that we are not defined solely by the sum of our corporeal parts, but also, and more importantly, by a dynamic configuration and kinetic rhythm particular to our individual composition (E IIL5). All individual modalities – whether human, building, or molecule – emerge from differential velocities and metastable tensions.[3] The singular plasticity of our motion-and-rest ensures our continuous self-modification, while retaining our individual perseverance. This is how Spinoza's concept of *modus* or modality is connected to his conception of motion and dynamic individuation. *Modus* in Latin has multiple meanings such as modality, manner, way, mood, rhythm, and musical scale. From Spinoza's perspective, individual modalities constitute ever-changing manners and moods of the cosmos. We are all different rhythms, unique ways of life.

But how does the process of individuation make way for an individual modality? Spinoza argues that once a number of bodies form close contact with one another, and sustain a shared rhythm particular to their collective composition, a new individual modality is formed as a composite manifold

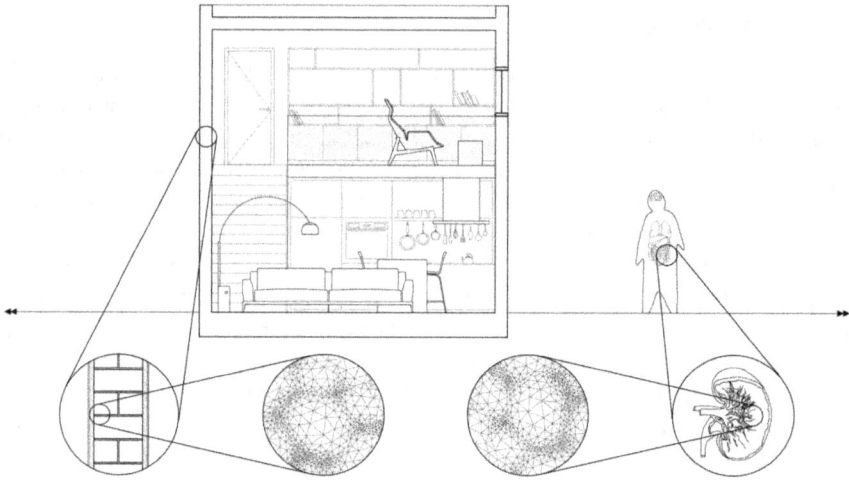

Figure 7.4 Emergent individuality

(E IIA2″). This means that every individual, through corporeal interlocking, is wrapped in other bodies that are in turn wrapped in other bodies *ad infinitum* (E IIL7S). A human body is itself a nested composition that functions at multiple organisational levels at once, composed of individual organs, which are in turn composed of individual cells, composed of molecules, atoms, subatomic particles, and infinitely small bodies. An architectural body is also a multi-scalar individual, composed of, say, individual bricks, which are in turn composed of individual clay aggregates, molecules, atoms, and once again, infinitely small bodies. Peculiarly, Spinoza's part-to-whole relationship draws parallels with contemporary emergence theory, which argues that the formation of complex individuals can be explained by interactions among simpler entities, insofar as they produce and sustain novel capabilities via their enmeshed configuration (Chalmers 2006: 244–6). That is, humans and buildings become unique individuals in their own right, to the extent that their singular configuration produces emergent capacities that their constituent individual parts such as bones, organs, and flesh, or walls, doors, and windows, do not exhibit by themselves.

In a letter to Henry Oldenburg (Ep. 32, 1665), Spinoza elaborates his part–whole relationship with a thought experiment concerning how lymph and chyle constitute blood:

> Now let us imagine, if you please, a tiny worm living in the blood, capable of distinguishing by sight the particles of the blood-lymph, etc. and of intelligently observing how each particle, on colliding with another, either rebounds or communicates some degree of its motion, and so forth. That worm would be living in the blood as we are living in our part of the universe, and it would regard each individual particle of the blood as a whole, not a part, and it could

have no idea as to how all the parts are controlled by the overall nature of the blood and compelled to mutual adaptation . . .

Spinoza argues that the part–whole relationship has nothing to do with the fixed order, harmony, or beauty of the cosmos. Rather, it functions via emergence and mutual adaptation, just as chyle and lymph constitute blood by engendering a third configuration through their rhythmic communication. What is usually forgotten in the analysis of this vivid example, however, is the significance of the worm. For Spinoza's tiny worm, despite its seemingly alien appearance, is itself another individual within the bloodstream composing this individual's bodily configuration. The presence of the worm implies that there is no self-contained individual made solely of its own essential, harmonious parts, but that every individual modality is always already a psychosomatic colony, a symbiotic combination of transversal habitats. Spinoza would have been delighted to learn the recent scientific discovery that human cells constitute less than half of the human body, while the majority is made up by cells of fungi, bacteria, microbes, and other non-humans (Yong 2016). He would have been thrilled to hear Jane Bennett's (2010: 112–13) example that even the crook of the elbow is 'a special ecosystem, a bountiful home to no fewer than six tribes of bacteria', which help 'moisturize the skin by processing the raw fats it produces'. This is Spinoza reminding us at the dawn of modern anthropocentrism that we have never been exclusively human. Rather, we have always co-evolved with and been co-constituted by the non-human, even in our own bodily configuration. And hence, every individual modality – whether human, animal, or building – is constitutively alien.

But how does individuality persist in time while undergoing constant modification? Spinoza's response is that individuals sustain their dynamic configuration via *conatus*, that is, the potence of an individual with which it endeavours to act and persist in its own being (E IIIP6–7). An individual's *conatus* is the persistency of expressing its singular rhythm, the plasticity of enduring internal fluctuations and surfing external oscillations, and the agency of acting upon life and enacting change. But then again, is *conatus* blind inertia, a reflexive drive, or a system of consciousness? In Spinoza's heterarchical ontology, *conatus* drives all individual modalities without exception. Humans, animals, buildings all act and persist for a finite duration with different degrees of power depending on their dynamic configuration, corporeal composition, and affective capabilities. While an architectural modality persists via structural integrity, conservation of momentum, and physico-chemical qualities, a human perseveres via operational metastability, feedback circles, and cognitive operations. So, each individual's *conatus* is a combination of singular powers and a unique range of affective capacities. Evan Thompson (2009: 85), a philosopher of cognitive science, sees Spinoza's *conatus* as a precedent to *autopoiesis*, which is theorised by biologists Humberto Maturana and Francisco Varela (1980: 78–84) as a capacity of living systems to structurally open themselves to incessant material flux, while simultaneously retaining their particular configuration through operational closures and feedbacks. What makes Spinoza's *conatus* even more radical, however, is that he recognises different degrees of *poiesis* (creative potence) in all individual modalities, and not only in living organisms. Spinoza's *conatus* makes it possible to acknowledge the capacity of each modality to exist, persevere for a finite duration, and affect its associated environment, while at the same time recognising their open-ended capabilities to differ and modify life in unique ways.

There is a strong reason why Spinoza's *conatus* emphasises both temporal perseverance and constant change. For Spinoza does not bifurcate life into subjects and nouns on the one side, and events and verbs on the other. Individual modalities are not rigidly bounded entities, upon which external events and situations fall. Rather, an individual perseveres, insofar as it retains its individual consistency, while simultaneously undergoing an infinity of individuating events. Modality and modification, individual and individuation, being and becoming are two expressions of one and the same thing for Spinoza. This coupling can be literally seen in his unfinished *Hebrew Grammar*, in which Spinoza draws attention to the fact that Hebrew verbs are not structured as distinct categories, but are derived from nouns via conjugation (CGH ch. 12–13). We usually overlook this double character of individuals, especially that of non-organic ones, and assume, for instance, that buildings are static and unchanging constructs, crystallised and frozen forms.[4] Yet as even the linguistic structure of the word 'building' in English (or *aedificatio* in Latin) strongly implies, a building is not only a consistent individual, a subject, a noun, but also a continuous individuation of building, an event, a verb.

Relational Affectivity

An individual modality never resides in a vacuum, but is always coupled, multiplied, and penetrated by a myriad of other modalities. What defines an individual, according to Spinoza, is not solely its characteristic rhythm, but also its affective capacities to interact with its environment. Spinoza elaborates these capacities via two concepts, namely, affection and affect, whose strategic roles should not be confused with one another. Affection (*affectio*) is a relational encounter in which the activity of an affecting modality is enveloped by and transcribed in the subjected modality as a trace (E IIP16). Suppose that it is deathly cold outside; you take shelter in a closed space, and become affected by the warmth of this interior. In this affective encounter, you record the intensive traces of temperature change, and translate its impact into your individual configuration. Affect (*affectus*), on the other hand, is the translation of these enfolded traces into an individual's transitive experience of power (E IIID3). Whereas affection is the connective line that puts modalities into interaction, affect is the individual passage from one mode of existence to another for a lived duration, resulting in an increase or decrease in that individual's power to be or to act – which are one and the same thing for Spinoza. Although affects are not strictly emotions, but pre-conscious fluctuations of our capacity of action as a result of modal interactions, Spinoza prefers to use conventional terms such as joy and sadness, by redefining them as transitions of power. In an affective encounter, an increase in

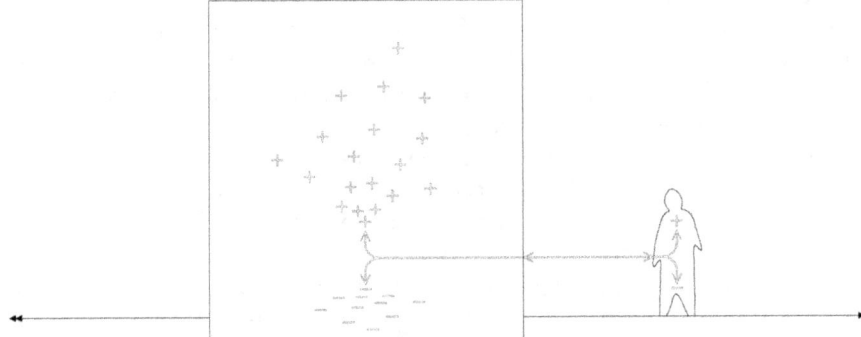

Figure 7.5 Affective coupling

an individual's power means an empowering modification; the affective cause of this empowerment becomes 'good' for the affected individual (E IVD1), and this empowering transition is translated as an affect of 'joy' (E IIIDef.Aff.II). By contrast, a decrease in power means a weakening modification; the affective cause of this disempowerment becomes 'bad' for the affected individual (E IVD2), and this disempowering transition is translated as an affect of 'sadness' (E IIIDef.Aff.III). So, there is no universal joy or sadness in Spinoza, but as many singular joys and sadnesses as there are trans-individual affections. Once you leave the cold behind and envelop the affective warmth of the closed space into your own modality, this encounter loosens the pressure on your metabolic thermoregulation and your power of being and acting becomes augmented. This is how your body translates a thermodynamic affection into an empowering affect of joy.

With this affective coupling, Spinoza starts to replace transcendent morality with immanent ethics. Morality is a system of judgemental values native to transcendent planes of existence. It functions via prescribed rules used for universally applicable distinctions of Good and Evil to constitute laws of command and obedience. Spinoza's ethics, on the contrary, corresponds to the open field of affective interactions. There is no pre-defined Good or Evil, but there are singular encounters that turn out to be good or bad, solely in relation to their empowering or weakening affects (E IVPref.). Salt water is not evil in itself, but for a freshwater fish it becomes a bad encounter for the simple reason of decomposing its power of existence. When Spinoza remarks that 'nobody as yet has determined the limits of the body's capabilities: that is, nobody as yet has learned from experience what the body can and cannot do' (E IIIP2S), he not only calls into question the supremacy of mind over body, but implies that the body's capabilities cannot be fully exhausted. We cannot know what a body can do in advance, since bodily capacities are actualised only in affective interaction with other bodies. I cannot know how deep I can dive or how much I can love, for they do not depend solely on myself, but they are emergent outcomes of my interactions and compositions with the sea and the depths, and with the individual I love and the affective networks in which we are enmeshed. Gilles Deleuze (1988c: 18–19) calls this discovery by Spinoza 'the unknown of the body', and deems it as profound as 'the unconscious of thought'. Individuals recognise their affective capabilities only through singular encounters, surprising all the parties involved by what their bodies can do, including themselves.

Rather than following pre-established rules, Spinoza's ethics is oriented towards experimenting with our affective capabilities and pushing our power to its limits. It is a dangerous ethos; one needs caution while experimenting with the unknown. Every time we open ourselves to an affective encounter, every time we take a leap, we risk weakening and even decomposing ourselves. Spinoza warns that 'we are in many respects at the mercy of external causes and are tossed about like the waves of the sea when driven by contrary winds, unsure of the outcome and of our fate' (E IIIP59S). This is why there is yet another turn in Spinoza's ethics, an orientation away from bondage of passions towards freedom of actions (E IVPref.).

Passions are passive affects caused by external affections that limit an individual's power of existence (E IIID1–3). We become subject to bondage if we surrender ourselves to the fluctuation of affective interactions. Instead, Spinoza suggests actively pursuing and constructing empowering compositions ourselves, by dwelling on active affects that spring from our own affective capabilities, by extracting joyful and empowering potentials from each encounter, and by doing justice to what life brings at every turn. This is an ethos of converting passions into actions. Passions result from our inadequate ideas, from confusing the real cause of our affective relations with their effects on our bodies. We can transform passions into actions, inadequate ideas into adequate ones, and constitute empowering compositions and joyful affects with other modalities, insofar as we discover an accordant rhythm, compose a common notion with their affective traces enveloped in our own bodies (E IIP37–9; E VP2–3). When you first encounter the sea, you absorb the kinetic rhythm and affective capabilities of the water through your body as a passive affect. Then, if you can find an accordant rhythm or compose an adequate commonality with the sea, you can transform the passive affect of your encounter into an active one. That is, you can then swim rather than be submerged, forming a joyful composition with aquatic forces.

According to Spinoza, this is our ethical journey in life; to collect as many affective commonalities with other modalities as possible, to transform passivising affects into activating ones, to evade and endure bad encounters while increasing and sustaining good ones, to affirm our power up to its very limits, and to increase the power of all modalities interwoven with ours. This is not a moral grid applied in advance and from above, but an ethical cartography drawn by each singular modality from within their unique voyage of life on the fly. In this ethical journey, Spinoza calls an individual with active affects a free individual (E IVP66S; TP 2.11). Spinoza's freedom, therefore, does not correspond to humanist notions of free will or free choice, but denotes an active and constitutive emancipation. Individuals are not born free, for freedom is not a static right or a possessable property, but a becoming; that is, we become free

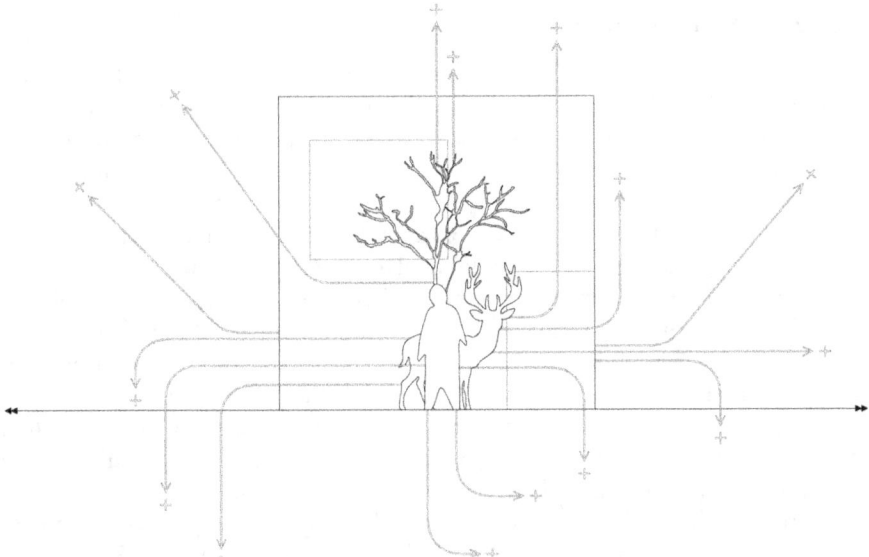

Figure 7.6 Ethical cartography

only by doing justice to our singular affective experiences via the combination of joyful experimentation and meticulous caution.

Spinoza's affect theory has lately witnessed a massive resurgence in diverse fields.[5] Antonio Damasio (2003: 11) and Heidi Ravven (2003: 261–4) go so far as to claim that Spinoza anticipated contemporary developments in affective neuroscience. What makes Spinoza's affective coupling even more radical, once again, is that affection and affect traverse all individuals, and are not exclusive to humans (E IIIPref.). That is, each and every modality – whether human, animal, plant, or building – enacts and undergoes affective modifications as a result of modal interactions. An earthquake becomes bad for a building insofar as it weakens or destroys the building's power of being. A flower enjoys the presence of a bee insofar as it helps augment the flower's reproductive continuity. This is not an anthropomorphic projection of human properties onto non-humans, for Spinoza suggests that it is the entire cosmos, not just a selected few, that expresses itself via affections or interactions, via affects or fluctuations of power. This implies that architectural modalities have singular affective capacities of their own. Such recognition prevents us from reducing buildings to passive backgrounds and neutral containers. From Spinoza's perspective, each architectural encounter becomes an engagement with vibrant modalities, with affective interactions that traverse us, with affects that increase or decrease our power, with experiences that take us over and transform us.[6]

Collective Enmeshment

According to John Colerus, one of Spinoza's earliest biographers, Spinoza was fascinated with spiders and their webs. When Spinoza wanted to divert himself from his studies, Colerus says, he took pleasure in watching spiders weave their webs and fight with each other to the extent that 'he wou'd sometimes break into laughter' (1706: 42). Watching how spiders enmesh with their milieu was presumably not a mere diversion for Spinoza, but an indicative empirical observation. For Spinoza suggests that individuals – whether human, spider, flower, or building – are not discrete or self-contained entities, but enmeshed modalities. Everything in nature is interrelated to other things (TIE 41). Individual modalities do not pre-exist their affective interactions and enmeshed compositions, but are constituted by their web of visible and invisible relations. These webs are not external to and distinct from individuals but a continuation of their modality. We all expand and contract ourselves by way of these enmeshed coalitions. We are all spiders weaving our milieu with affective webs of our own.

Defining individuals as enmeshed modalities has immense consequences, especially for technological, sociopolitical, and ecological meshworks. Spinoza suggests that if two or more individuals share compatible channels of dynamic configuration and affective communication, they might as well constitute a third enmeshed modality for the duration of their coupling (E IIA2"D). This means whenever we are entangled with technological meshworks, we expand our modality to the affective limits of our enmeshed composition. The drawing of an architectural design sketch does not solely consist of the external relations among discrete individuals such as an architect, a drawing table, a pen, and paper. The act of design is constituted by expanded meshworks of architect-and-pen, pen-and-paper, paper-and-table, or rather, a distributed meshwork of

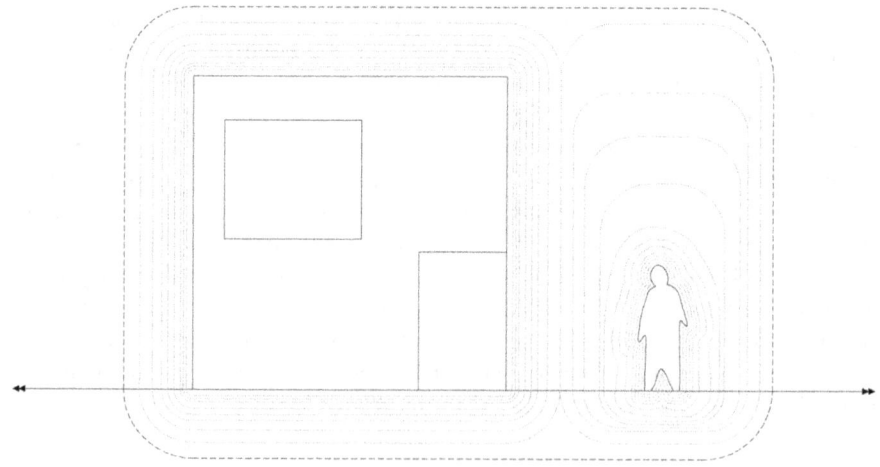

Figure 7.7 Technological enmeshment

architect-and-pen-and-paper-and-table. In turn, this modality is enmeshed with other architects, computers, and office spaces, which are entangled with architectural discourses, historical precedents, and technical capabilities, which are interwoven with other experts from engineering, construction, and landscape, which are enrolled in professional bodies, economic networks, and bureaucratic apparatuses, among an infinity of other enmeshments. This means that we do not use tools as external gadgets, but rather multiply our modality with and through them.[7] From Spinoza's perspective, architecture is our non-human becoming, our prosthetic web of affective relations between our body and milieu. A building is no longer a separate receptacle but a topological extension of our modes of existence. Together, we constitute bio-technological meshworks with emergent capabilities in each instance.

Sociopolitical enmeshment constitutes another focal point throughout Spinoza's oeuvre. He goes so far as to define the ethos of an individual not solely as increasing its own power, but simultaneously as furthering the power of all modalities via sociopolitical enmeshment. In the *Ethics*, Spinoza suggests that people 'all together should endeavour as best as they can to preserve their own being, and that all together they should aim at the common advantage of all' (E IVP18S). But why should individuals pursue the politics of collective empowerment, rather than the politics of self-interest? In his unfinished *Political Treatise*, Spinoza suggests that if individuals were to act solely from self-interest and the masses (*vulgus*) do just as they please, there would constantly arise clashes of interest and unnecessary struggles, through which hatred, anger, and deceit would prevail, as witnessed repeatedly throughout history (TP 2.13–14). Instead, Spinoza envisions an absolute democracy immanently constructed by a constituent multitude, who share their natural rights in common while retaining their singular differences, and enhance the wellbeing of one another while constantly renewing the reciprocal configuration of their social enmeshment (TP 11.1–3).[8]

Spinoza's politics was in clear opposition to prevalent political theories of his time, especially that of Thomas Hobbes. Hobbes (2003 [1642]: 3–4) argued that 'man is a wolf to man', that people can never solve their clashes of interest by themselves, so they need to be kept in constant check and disciplined by an absolute state. As a result, Hobbes (1998 [1651]) theorised a transcendent state

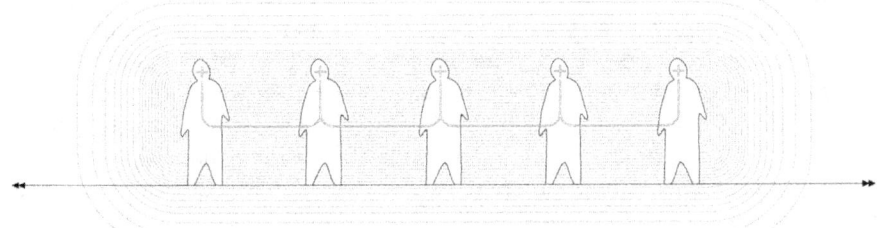

Figure 7.8 Sociopolitical enmeshment

apparatus, reigning above its citizens, who were to transfer almost all of their decision-making power to the state's central authority via the social contract. Spinoza, on the contrary, conceived that 'man is a God to man' (E IVP35S), that people can constitute the most empowering social meshworks by collective self-organisation, so that the state's role is primarily to guarantee democratic freedom and promote reciprocal empowerment. In Spinoza's own words (TTP ch. 20/G III 241):

> [The state's] ultimate purpose is not to dominate or control people by fear or subject them to the authority of another. On the contrary, its aim is to free everyone from fear so that they may live in security so far as possible . . . It is not, I contend, the purpose of the state to turn people from rational beings into beasts or automata, but rather to allow their minds and bodies to develop in their own ways . . . Therefore, the true purpose of the state is in fact freedom.

Throughout the Enlightenment, Spinoza's political theory was continuously suppressed, and it has only recently witnessed an extensive resurgence. Post-war political theorists such as Antonio Negri (1991), Louis Althusser (2006), and Etienne Balibar (2008) found in Spinoza the theoretical equivalent of the self-organised desires of May '68, and attempted accordingly to realign revolutionary theory and practice away from party-based organisations and representative institutions towards bottom-up collectives and constitutive enmeshments. In architecture, Spinoza's opposition to the subordination of the decision-making powers of everyday social actors hints at a confluent trajectory that challenges the hegemonic roles of clients and economic modes of production, state apparatuses and top-down planning initiatives, architects and corresponding experts, which tend to reduce the role of everyday users in the built environment to 'passively experienc[ing]' whatever is 'imposed upon them' (Lefebvre 1991: 43). From Spinoza's sociopolitical perspective, alternative experiments become viable, which affirm the constitutive agency of everyday users in shaping their built environment, not instead of but enmeshed with all the relevant spatial actors.

Ecological enmeshment, finally, is what brings all meshworks together. Spinoza's heterarchical plane knows no real distinction between nature and culture, recognises no supremacy of culture over nature. Spinoza explicitly warns against seeing ourselves as distinct from and above nature, and argues instead that we exist as an affective part of nature's modal configurations (E IIIPref.). *Contra* cultural supremacists who constantly widen the gap between nature and culture to legitimise their instrumentalisation of nature, Spinoza advocates that there is no chasm between nature and culture in the first place. This was Spinoza's early warning at the dawn of the Enlightenment against the narcissistic tropes of modern anthropocentrism that promoted humanity's conquest of nature, a warning that has influenced diverse circles from nineteenth-century Romanticism to post-war ecological movements. For Spinoza, nature and culture constitute a continuum. We live as part of this common ecology that equally houses all that are distinguished as human and non-human, natural and artificial, urban and non-urban.

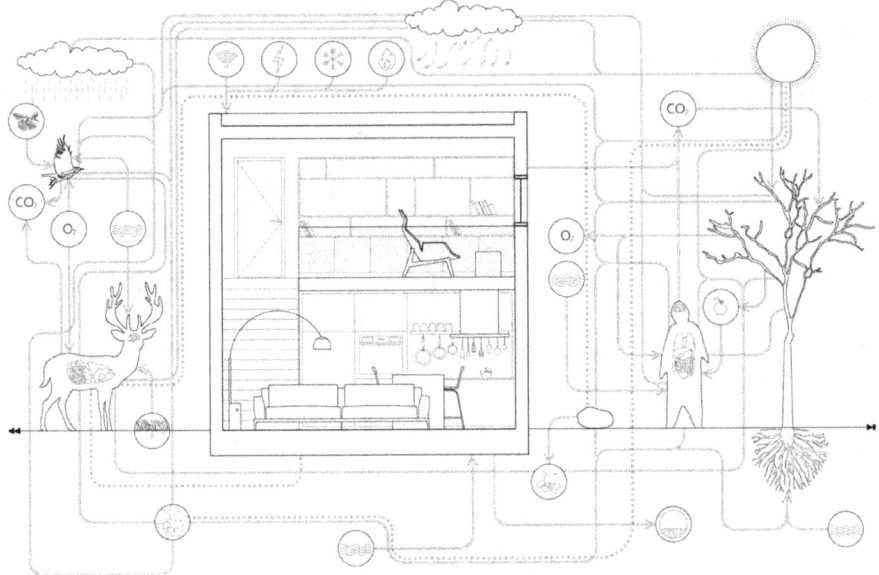

Figure 7.9 Ecological enmeshment

Individual modalities – whether planets, humans, computer software, or buildings – exist in entanglement with their associated ecological meshwork. This entanglement is extensively covered in the phenomenological philosophy of Edmund Husserl, Martin Heidegger, and Maurice Merleau-Ponty. Although their focus on the existential relationship between individuals and their milieu was a welcome contribution after dominant forms of mind- and subject-centred understanding of individuals for centuries, there still arises a mismatch with Spinoza's system because of the exclusivity of human subjects in orthodox phenomenology, as the privileged monads experiencing their being-in-the-world before all others. Jakob von Uexküll, an eccentric Estonian biologist who is deemed a latent Spinozist by Deleuze, shows us another dimension when he elaborates the relation of individuals with their milieu through the world of animals. One of Uexküll's (2010: 41–3) most notable contributions to theoretical biology is his notion of *Umwelt*, defined as an associated environment that corresponds to the singular affective web of any given organism. This is not a reiteration of the simple fact that each individual exists within a milieu. Uexküll shows, additionally, how each animal creates its own affective meshwork via sense and effect cycles, and hence weaves its own selective and singular milieu by augmenting certain relations, attending less to others, while fully ignoring most connections as background noise. This means there are as many different worlds of experience as there are different animals. With Uexküll's emphasis, we start paying more attention to how ticks, sea urchins, dogs, amoebae, and jellyfish experience their world and constitute their own ecological enmeshment.

From Spinoza's ecological lens, however, we are still missing yet another dimension, that of non-organic modalities. For Spinoza does not profess being-in-the-world to be a distinctive trait of humans or animals, but suggests that all modalities create their own affective milieu; each modality is a singular becoming-of-the-cosmos itself. This opens up all kinds of questions, especially for architecture. What is the singular *Umwelt* of an architectural modality? What are the peculiar ecological enmeshments of a building? How does an architectural construct constitute its associated milieu? Today, the critical question of ecology in architecture can be addressed neither by the anthropocentric vanity that still holds on to the narrative of mastery over nature, nor by nostalgic illusions that sublimate nature to a divine Mother Earth, nor by cosmetic exploitations that market greenwashing strategies as ecological sensibility, nor by quantitative codifications that reduce ecological interaction to earning points and satisfying the prerequisites for eco-friendly certifications. Spinoza's ecological approach hints at a different trajectory that starts from acknowledging our shared continuum with nature, and paves the way for undertaking symbiotic experiments that reciprocally empower all modalities within this nature–culture continuum.[9]

These are the first steps towards unfolding the generative potentials between Spinoza's philosophy and architecture. One must be cautious, however, not to apply Spinoza to architecture as a transcendent framework, not to turn his philosophy into a pre-fixed method, not to refer to the same immutable codes each time these two planes confront each other. The challenge, rather, is how to come up with novel modifications of both planes at once, how to unsettle problematic presuppositions of architecture and philosophy together, and how to push these experiments to their limits by doing justice to each singular interaction. A risky voyage to subterranean passages without pre-inscribed guarantees but full of magmatic surprises. It all starts with the affirmation of a modest yet missed encounter, to be explored via infinitive verbs and question marks, rather than premature conjugations and full stops.

Notes

1. See, for example, the chapters by Rawes, White, and Frichot in this volume.
2. Spinoza's elaboration of modalities in the *Ethics* is much more sophisticated than can be expressed in this compact summary, including different categories such as infinite and finite modalities, as well as mediate and immediate modalities. In this chapter, we will focus mostly on finite modalities.
3. What Spinoza means by 'individual' is still a topic of debate among Spinoza scholars. For example, Bennett (1984: 107) thinks that Spinoza reserves the term 'individual' for living organisms, whereas Garrett (1994: 90) suggests that, for Spinoza, non-living things also count as individuals.
4. See Kwinter (2002) for a rigorous theoretical treatise defining architecture as event.
5. For different aspects of contemporary affect theory, see Massumi (2002), Thrift (2007), Protevi (2009), and Gregg and Seigworth (2010).
6. See 'Fun Palace' by Cedric Price and Joan Littlewood (1968) for a rare architectural design project that acknowledges the affective agency of buildings and investigates

how architecture can both undergo and instigate constant change with the help of user potentiation and cybernetic systems.
7. For contemporary theories that recognise our enmeshed relationship with technology and architecture as a direct extension of our own modalities, see Bateson (1987: 256, 323–6), Deleuze and Guattari (2005: 71–4, 89–90), Teyssot (1994: 15–16), and Mitchell (2004: 1–41).
8. See Negri (1991: 144–210) for a complementary interpretation of how Spinoza developed his notion of absolute democracy and multitude.
9. See Grosz (2001) for an architectural theory that starts from the ecological acknowledgement of a shared nature–culture continuum.

8

Dissimilarity:
Spinoza's Ethical Ratios and Housing Welfare

Peg Rawes

As this collection shows, ideas linking *ratio* (or reason) to human qualities are present throughout Spinoza's writings. In the *Ethics* these relationships are described through a geometric method in which equality is constructed, counter-intuitively, from the uniquely differentiated God/Nature. Human nature is also uniquely differentiated into the mind and the body. Yet this *dissimilarity* between the attributes also defines a relationship between them. Together, within the person, mind and body compose a kind of correspondence or equality of attributes. Human life is therefore formed through a mutual recognition of difference in each 'element' (for example, our mind or body) rather than a definition of individuation that is based upon finding similarity between mental or physical capacities. In the long closing scholium to Part II, Spinoza argues that his doctrine of these relations has 'practical advantages' for understanding human life. Interestingly for this chapter, which considers the relationship between the *Ethics* and housing welfare, he also states that society is assisted by this doctrine of relations which 'teaches us to hate no one, despise no one, ridicule no one, be angry with no one, envy no one. Then again, it teaches us that each should be content with what he has and should help his neighbour . . . solely from the guidance of reason as occasion and circumstance require' (E IIP49S).

For Spinoza, reasoning is not confined to our intellectual powers, which have tended to be defined as independent from a specific corporeal body or societal value (for example, as in mathematical or geometric reasoning). Rather, the *Ethics* shows that the power of reasoning is intimately connected to differentiated and specific human modes of expression, especially our affects and emotions, which are often considered to be most strongly affiliated with corporeal and irrational expression. Instead, for Spinoza, reasoning is a capacity that is directly related to our affective corporeal capacities for producing ethical individuals (and society). Reasoning, we find, is a consequence of these transitive and relational qualities, which rationalist thinking generally seeks to exclude for their contravention of the constancy associated with rational and intellectual thought.

Interestingly, however, Spinoza locates his thesis within a process of geometric reasoning first used by Euclid. Unlike traditional scientific or mathematical understandings of geometry, forms of intellectual reasoning that are often explicitly disconnected from the irrationality and inconstancy of daily life, Spinoza shows that geometry is in fact deeply connected to it. Indeed, through his discussions of geometric subjects, for example points and lines, and in the manner that he uses the geometric method to organise the text, his aim is to reveal a '*different standard* of truth' (E IIApp., my italics).

Below I show how this difference, or rational dissimilarity, is at the core of Spinoza's argument, and leads to his discussion about the differences between nature, mind and body, the affects, geometric thinking, and hence, societal value. In each case, Spinoza shows that the individuation of an entity is wholly distinct, but also located in a ratio of differences with other entities, including other minds and emotions. I close the first part of the chapter by suggesting that the *conatus* is the most significant expression of this ethical dissimilarity, the 'site' of our wellbeing, and hence, our agency to design our personal and social welfare. I refer to Spinoza's discussion about the design of a house as an expression of our powers of wellbeing, drawing from the ratio between the imagination and the desire to inhabit a home.

Having explored Spinoza's conception of dissimilar ratios, I then consider the social and architectural relations that composed seventeenth-century housing welfare in the Dutch Republic, specifically the almshouse tradition (which has interesting correspondences with the English almshouse tradition during this period). While Spinoza's discussion of architecture and housing is not explicitly concerned with specific social or urban concerns of the Dutch Republic, there are a number of connections between 'ethical ratios' and almshousing; for example, almshouses are composed of varying ratios of individuation and collective affectivity insofar as they provide housing (a home) for individuals who have inadequate economic or societal wellbeing, and an agency that is more acted on by these forces. Almshouses represent a kind of social-spatial ratio of housing and wellbeing that benefits individuals and society overall, because residents gain the dignity of a home, as opposed to the indignity of homelessness. Almshouse ratios of home and dignity therefore suggest a more equitable way of living together *with* our differences and dissimilarities.

Finally, I consider how Spinoza's ratios of dissimilarity can be used to discuss the contemporary housing situation in the UK. The lack of access to good-quality and affordable housing for many families and households in the UK is currently producing a housing crisis which is the result of complex historical, political, and social issues over the past forty years. Taking note of this complexity, I briefly consider some aspects of economic and spatial dissimilarity to suggest that these contemporary questions about inequality in housing can be discussed through the principle of dissimilar ratio that I find in the *Ethics*. While Spinoza's aim is to present the positive benefits of rational thinking for a more equal society, he also provides a fascinating analysis of the negative qualities that generate current unequal ratios; for example, the impact of the passive affects of envy, jealousy, and melancholy. Thus, while dissimilarity is a positive interpretation of different spatial relationships in almshousing, I also show that unequal ratios in housing design can be understood with reference to Spinoza's discussion of passive affects. In particular, the dissimilarity or imbalance between the economic, spatial, and wellbeing ratios in current housing provision represents an especially negative dissimilar ratio of housing and dignity for many families in the present day.

Dissimilarity

Dissimilarity between attributes, modes, and affects is a key property of reasoning in the *Ethics*. From the text's opening definitions Spinoza shows that difference defines the uniqueness of God, substance, and all life that follows. Difference extends from the indivisibility of God or substance, through the differently individuated modes of mind and body and their affects, to the beneficial societal tolerance of human differences (E IIP49S). Definition 3 of Part I of the *Ethics* states 'By substance I mean that which is in itself and is conceived through itself; that the conception of which does not require the conception of another thing from which it has to be formed.' This is followed by a series of propositions on the importance of difference for a conception of substance, for example: 'Two substances having different attributes have nothing in common' (E IP2), or 'When things have nothing in common, one cannot be the cause of the other' (E IP3), or 'One substance cannot be produced by another substance' (E IP6).

However, this emphasis on dissimilarity also comes with the statement that bodies and thoughts are produced *in relation* to other bodies and minds; for example,

> A thing can be said to be finite in its own kind . . . when it can be limited by another thing of the same nature . . . A body can be said to be finite because we can always conceive of another body greater than it. So too a thought is limited by another thought. (E ID2)

Later, when discussing the affects, Spinoza states: 'The human mind has no knowledge of the body, nor does it know it to exist, except through ideas of the affections by which the body is affected' (E IIP19). Thus, for human beings, the relationship generated between our different mental and physical conditions results from our affects. These are the powers that connect the entirely distinct modes of mind and body. In Part III Spinoza explains their relational value for producing activity or passivity in the individual, because 'The human body can be affected in many ways by which its power of activity is increased or diminished' (E IIIPost.1). Affects are therefore ideas that produce a mental and bodily correspondence, or ratio, between modes of different attributes.

Spinoza deploys Euclid's classical *geometrico ordinare* as a textual method through which he examines the varied mental and physical natures of the individual. In the first three parts of the text, Spinoza follows Euclid's method, redeploying the geometric 'elements' – the axioms, definitions, corollaries, propositions, and scholia – to demonstrate that being human is inherently related to the powers of nature and of God. On the face of it, each mathematical 'function' appears to be directed towards a rational form of argument. However, Spinoza's method is clearly not pure 'rationalism' in the traditional scientific understanding of a consistently repeatable demonstration. The formality of the geometric method breaks down regularly with long interruptions or asides, and the proportion of length and quantity between the various geometric elements is inconsistent. By the time the text gets to Part V, its precise geometric form has broken down almost completely. Most strikingly, of course, the subject matter is clearly not a standardly rational 'matter'. The irregular and inconsistent 'elements' therefore compose a geometric *rationale* that is constituted from dissimilarity, rather than a logic of harmonious and symmetrical forms (for example, recurring in the associations between geometry and symmetry in architectural design). Spinoza's approach also reminds us that ratios in geometric reasoning are composed of two distinct entities put into relation with each other: for example, when written in algebra, geometric *figures* are composed of multiple functions (for example, the ratio of a hypotenuse is written as $b/c = \sin(b)$), showing that the unity or idea of a geometric figure *itself* contains a dissimilarity in its form and structure. Contemporary forms of architectural computational design, especially 'parametric' software programmes, actively use these differential ratios in the design of complex geometric (and often organically inspired) structures. Geometric ratio is always the description of more than one entity, and we might therefore say that reasoning, for Spinoza, is always a relational practice.

The *conatus* and Wellbeing

Spinoza shows that the most important site of this dynamic relationship is located in the *conatus*: that '[which] endeavours to persist in its own being but which is nothing but the actual essence of the thing itself' (E IIIP7). An energetic ratio or reasoning composes the *conatus*, or life, and so our capacity for agency and self-determination is shown to exist precisely within the interaction between our positive and negative affects. Spinoza uses much of Part III to explain how we are variously affected by different things, both positively and negatively. This capacity to be affected is not reduced to a universally similar form. In *Ethics* IIIP51 his explanation of the capacity for dissimilarity of individuation is explicit: 'Different men can be affected in different ways by one and the same object, and one and the same man can be affected by one and the same object in different ways at different times.' When more pleasurable, these affects can lead us 'to a state of greater perfection', but when painful, they reduce us to a state of 'less perfection' (E IIIDef.Aff.II–III). Thus, we find that the affective powers that constitute our *conatus* are another way of showing that our subjecthood or agency arises out of a dynamic relationship between positive and negative corporeal powers: relations that are distinctly human modes of differentiation.

In the Appendix to Part I, Spinoza writes of our capacity to be affected and our consequent sense of 'wellbeing', which is produced in two ways: first, from our imagination's power to generate aesthetic agreement or disagreement between our ideas and nature (e.g. 'those who mistake their imagination for intellect', E IApp.), and secondly, from the stimulation of our nervous system by our senses (including sounds, smells, tastes) that 'are conducive [or not] to our feeling of well-being' (E IApp.). Wellbeing is therefore both a mental and a corporeal experience of an agreement between entities, which is associated with harmony or beauty, but which can also be produced out of the agreement between our imagination and the 'nature of things'. However, he continues by observing that there are many variances in what stimulates aesthetic or sensory wellbeing:

> For although human bodies agree in many respects, there are very many differences, and so one man thinks good what another thinks bad ... all of which show clearly that men's judgment is a function of the disposition of the brain, and they are guided by their imagination rather than intellect. (E IApp.)

Our capacity for wellbeing therefore results from diverse affective relations; mental and physical, internally and externally derived, affirmative and negative. In addition, Spinoza's close study of the negative emotions and their affective powers, such as jealousy and melancholy, highlights that our wellbeing is not an idealistic or constant state. Emotional and physical pain or discomfort impede our personal wellbeing and are also negative for society: 'That which so disposes the human body that it can be affected in more ways, or which renders it capable for affecting external bodies in more ways, is advantageous to man ... On the other hand, that which renders the body less capable in these respects is harmful' (E IVP38), or 'Whatever is conducive to man's social organisation, or causes men to live in harmony, is advantageous, while those things that introduce discord into the state are bad' (E IVP40).

Mental and physical wellbeing also inform Spinoza's discussion about the design of a house in Part IV. Here he shows how the affects and imagination combine in the *conatus* of the person who is designing a house, which is expressed as an aesthetic experience. Spinoza explains how the ratio between the individual's 'appetite' for a house (a home: somewhere for the *conatus* to live, expressly) and the design process constitute a reflexive state:

> when we say that being a place of habitation was the final cause of this or that house, we surely mean no more than this, that a man, from thinking of the advantages of domestic life, had an urge to build a house. Therefore, the need for a habitation in so far as it is considered a final cause is nothing but this particular urge. (E IVPref.)

In this architectural reflection, Spinoza suggests that housing design can be achieved from a ratio between an imaginary and a physical pleasure, rather than viewing architectural design as being generated by the implementation of preceding discrete or idealised Platonic forms or pre-determined plans.

In this respect, the capacity for wellbeing is reflected in the desire to design a place in which the individual (*conatus*) can thrive; that is, the home. Spinoza considers the different material, physical, and aesthetic powers of expression that are brought into ratio with each other in the design of a house. We might go so far as to say that to house the self well is an aesthetic form of 'care for the self' or of a kind of 'housing welfare'. At this point, it is worth noting that Spinoza's form of geometric thinking contrasts with modern, rationalist architectural designs. Rationalist architecture has often been seen to equate with idealised geometric, economic, and rational principles that then undermine individual freedoms and differences, and are associated with socially deterministic or technocratic management systems. When applied to large-scale social design projects, especially housing, these have had a particularly controversial history and value.[1] However, this chapter considers how the *Ethics* might help to discuss questions of wellbeing and housing, and so ratio here is not just a concern with economic or design efficiency, but may in fact be the basis for understanding ethics through a principle of dissimilarity.

Early Modern Dutch Almshouses

During Spinoza's lifetime almshouses were an important form of early modern housing welfare and social care in the Dutch Republic's fast-growing and wealthy cities.[2] Social and economic historical research by British and Dutch scholars has shown that there were close correspondences between Dutch and English almshouses up to the end of the seventeenth century. Historians Nigel Goose and Henk Looijesteijn define almshouses as:

> those institutions specifically designed to provide accommodation for the elderly poor, whether established by an ecclesiastical foundation, a monarch or member of the ecclesiastical hierarchy, by a company or by a private individual of whatever status. Their purpose was to provide free, sheltered accommodation for elderly people of modest means, usually in discrete apartments within a larger edifice, which may or may not have been supplemented by additional benefits in the form of a weekly stipend and allotments of food, clothing or fuel. While local authorities might sometimes have intervened in the life of these institutions, and the state may have become increasingly involved in their regulation, their defining characteristic is their provenance in the realm of philanthropy, whether founded by a private person or by an organization. (Goose and Looijesteijn 2012: 1052)

Even in times of harsh social attitudes towards the poor (members of society made passive through lack of economic or other forms of agency), almshouses represent 'a degree of continuity' of 'concern to provide succour to the impotent, elderly (and usually local) poor – across the divide created by both humanist social theory and the Reformation – even as attitudes towards the idle and the dissolute hardened' (Goose and Looijesteijn 2012: 1050). Begun in the fourteenth century 'as a way of caring for the elderly poor' (Buursma 2011: 115), almshouses were the main form of charitable social housing and social care until the late nineteenth century, when housing associations became the leading form of institution in both Dutch and English contexts (Goose and Looijesteijn 2012: 1051).[3] During the seventeenth century, the urban expansion of the wealthy central and western cities of the Dutch Republic, especially in the Province of Holland (including Amsterdam, Haarlem, Leiden, and The Hague), included the establishment of 'small courts', 'hofjes', or 'hofs' (Goose and Looijesteijn 2012: 1053). Run by groups of trustees, these almshouses were charitable institutions endowed to a parish by wealthy businessmen and women on their death to provide free housing, sometimes with financial welfare ('dole', previously called 'alms'), to single women and men over the age of 50 on low incomes.[4]

Goose and Looijesteijn also observe similarities in the Dutch and English almshouse traditions; in the design and architecture, the principles of care and organisation, and in the social status of occupants and founders. However, one key difference between the two societies is in the funding provision, with the Dutch Republic reliant upon non-statutory charitable donations from private individuals for its poor, while the English Poor Law tradition established statutory poor relief through taxation (Goose and Looijesteijn 2012: 1049–52). Nevertheless, in the context of the increasingly urbanised and socially unequal Dutch Republic, almshouses represented an important charitable way of designing wellbeing into housing.

In The Hague, where Spinoza lived between 1671 and 1677, and where he completed the *Ethics*, approximately 115 almshouses are recorded as having been built during the seventeenth century (Buursma 2011: 122).[5] Usually they were composed of small dwellings around a courtyard, a *liefdadigheidshofje* or 'almshouse-in-court' (Goose and Looijesteijn 2012: 1053). Smaller almshouses were formed of a row or terrace of houses. One of the courtyard almshouses, Helige Geesthofje, is almost directly opposite number 72–74 Paviljoensgracht, the house where Spinoza lodged and where he completed the *Ethics*. Founded in 1616 by the wealthy Protestant institution the Helige Geestmeesters, this hofje is still an active residence today, managed by the traditional almshouse mistress or *binnenmoeder* (Buursma 2011: 126). Composed of a courtyard with thirty-seven dwellings, Spinoza would have passed it on a regular, even daily, basis.

Just a few streets away at 49–85 Lange Beestenmarkt is another seventeenth-century courtyard hofje, Hof van Wouw, which again Spinoza might have known about. Hof van Wouw still functions as an almshouse for women over 50 (now of mixed denominations). It was founded by Cornelia van Wouw, a local businesswoman from a regent family in The Hague. Unusually, however, the hofje was founded during Cornelia's middle age in 1647 rather than after her death. Although she would certainly have established the house as a means to improve her own personal and social stature (while alive and in the afterlife), she lived in the front gatehouse until she died in 1671, and would therefore, presumably, have had the *binnenmoeder* responsibility for her residents.

Architecturally, the hofje has a number of design qualities that reflect the ratios of wellbeing in Spinoza's thinking. The designer of Hof van Wouw is not known with certainty, although the architectural historian Gonda Buursma suggests that it might have been 'the town architect, Bartholomeus van Bassen' (2011: 122). Eighteen cottages are set around a 'picturesque' (Buursma 2011:

SPINOZA'S ETHICAL RATIOS AND HOUSING WELFARE 115

Figure 8.1 72–74 Paviljoensgracht, The Hague (photo: the author 2015)

122) courtyard, each composed of a ground floor single room and upper level. Restored twice in the twentieth century, the cottages provided the women with private internal domestic space, and a shared courtyard space away from the public urban sphere. While it has been considerably renovated since the

Figure 8.2 Hof van Wouw courtyard (photo: the author 2015)

seventeenth century, it features architectural elements and design which have an affinity with the principles of aesthetic agreement that Spinoza observes to be characteristic of wellbeing. The hofje's restored architectural features include its red-brick facades and stable doors, pediments above each of the window gables, capped chimneys, and a pine cone, a sign of hospitality, above the entrance gate. Restorations have reinstated a seventeenth-century formal geometric garden, a well, and a pathed, bordered garden with fruit trees, which Buursma suggests are derived from Cornelia's instructions (2011: 122). Added at a later date, a passageway leads from the front garden into a rear garden, again laid out in a seventeenth-century geometric style, and which is actively gardened today. Now providing fourteen houses at a very modest rent for the centre of The Hague,[6] Hof van Wouw still operates under its original premise of providing a housing community for women of modest income, as when it was established in the seventeenth century. Architecturally, historically, and in a contemporary urban context, the hofje suggests how the almshouse tradition can cultivate varied social, physical, and aesthetic qualities that enhance its inhabitants' powers of wellbeing to flourish in a manner not unlike the way in which Spinoza presents his rationale for differentiated ratios enabling the wellbeing of the *conatus*.

This hofje was used 'as a model' for the largest and most spacious of The Hague's seventeenth-century courtyard almshouses (Buursma 2011: 121). Comprising sixty-two houses with two storeys per house and an attic, the Hofje

Figure 8.3 Hof van Wouw garden (photo: the author 2015)

van Nieukoop at 36–202 Warmoezierstraat (Goose and Looijesteijn 2012: 1658–62) was designed for businessman Johan de Bruijn van Buijtenwech and his wife Maria Cornelia van Duyveldandt by one of the most important architects of the period and region, Pieter Post (Buursma 2011: 121). However, unlike the two courtyard almshouses mentioned above, the Nieukoop is now an example of contemporary private market housing trends; in contrast to its previous provision of charitable housing, it is now highly sought after in the private rental sector (although its properties are leased at a reduced rate, given its historic significance and central location in the city).

Bringing together Spinoza's rational analysis of human wellbeing with seventeenth-century housing welfare, we can see that both are concerned with enabling the agency of the individual within society. Spinoza shows how our different affective powers combine in the *conatus*'s capacity for wellbeing, composing a care of the self. The almshouse represents a form of social housing in which residents' mental and physical wellbeing is encouraged to flourish socially within the community, in contact with nature, and reflected in the design of a hospitable architectural environment. In addition, both Spinoza and the almshouse tradition suggest that an increase in a person's dignity or wellbeing also benefits society. Although it could be objected that almshouses were partly motivated by the desire to remove the poor from the public realm, less economically powerful members of society were given housing off the street, while their wealthy Dutch benefactors were thought to gain spiritual improvement in the afterlife. While there has been extensive scholarship on the economic inequality of the Dutch Republic during this period, both Spinoza and the almshouse tradition suggest that mutually beneficial spiritual and physical flourishing also contributes to the prosperity of society: 'Whatever is conducive to man's social organisation, or causes men to live in harmony, is advantageous' (E IVP40).

In the next section, I consider twentieth-century and contemporary UK social housing examples of wellbeing and housing ratios. I begin, however, by noting that these ratios are currently characterised by poor-quality and limited affordable housing and stark inequality in access to housing, rather than equality or wellbeing. A principle of dissimilarity, derived from Spinoza, is therefore a helpful critical tool for discussing the increasingly disproportionate ratios that define the wellbeing and spatial capacities of social and affordable housing, together with the alarming rates of inequality that are experienced across social groups, including the young, the old, and families who have reduced economic power or agency in the UK. I close the chapter on a more positive note, with a brief discussion of examples of housing in which wellbeing is positively designed into ratios of architectural space, environment, and community.

The Contemporary UK Housing Crisis

Spinoza's observation that passive or negative affects detrimentally impact upon an individual's wellbeing is an explicit issue in the current UK housing context. Researchers specialising in inequality and housing, such as the charities Shelter and the Joseph Rowntree Foundation, have observed alarming increases in mental and physical health-related problems in UK households, including asthma and depression, which they conclude are partly a result of overcrowded and poor-quality housing (Shelter 2013b; National Centre for Social Research and Shelter 2013; Harker 2006; Joseph Rowntree Foundation 2015). The Royal Institute of British Architects' 2011 *Case for Space* report cites a figure of £21,815,546 as the annual cost to the NHS of overcrowding resulting from poor housing (RIBA 2011: 13). First published in *The Real Cost of Poor Housing* (2010) by the Building Research Establishment, this research also observes that

> the total cost [to society of poor housing in England] is some £600 million per year in terms of savings in the first year of treatment costs to the NHS if these hazards were removed, or at least reduced to an acceptable level. The full costs to society are estimated to be some £1.5 billion per year. (Roys et al. 2010: 11)

In 2005, the Shelter report *Full House? How Overcrowded Housing Affects Families* observed that overcrowding harms family relationships, with 92 per cent of survey respondents stating that lack of privacy was a key concern for wellbeing, and 81 per cent stating that overcrowding caused fighting and arguing among children (Reynolds and Robinson 2005: 8; Harker 2006).

Studies in the past decade, including Shelter's *Little Boxes, Fewer Homes:*

Setting Housing Space Standards Will Get More Homes Built (2013) and studies from RIBA, draw attention to the impact of declining living standards because of small and poorly designed new affordable housing in the UK. For example, the average newly built home in England is 92 per cent of the Greater London Authority's 2010 space standards' recommended minimum size, which equates to the loss of space in a one-bedroom home for a single bed and accompanying furniture (equivalent to a three-seat sofa, desk, and chair). In a three-bedroom home of 76 sq m, this means the loss of an entire single bedroom and its contents, or a 'galley kitchen' and small table (RIBA 2011: 5). New housing in Britain is also smaller than housing built in western Europe, with the average new home measuring 76 sq m and composed of 4.8 rooms. In comparison, Ireland's new homes are 15 per cent bigger than UK housing (87.7 sq m), Dutch new housing is 53 per cent bigger (115.5 sq m), and Danish new builds are 80 per cent bigger (137 sq m) (RIBA 2011: 10). Secondly, UK housing is sold per room (rather than by square metre as in Europe and North America) so that prospective buyers are often 'poorly-equipped' (Shelter 2013b: 5) to properly scrutinise the size of rooms or the total size of a house. Shelter notes that developers have been reported to use marketing techniques, including installing reduced-sized furniture, to disguise the scale of rooms (Shelter 2013b: 5).

These recent studies highlight that historically, and in contemporary housing provision, reductions in space standards detrimentally affect individual and societal wellbeing. If we consider these housing issues in light of the notion of ratio found in the *Ethics*, we can say that overcrowded houses produce negative affects which depress the agency and wellbeing of the individual. The negative impacts of poor standards of wellbeing in affordable housing combine with decades of decline in building the necessary amount of affordable and social housing for the UK population. In *Solutions to the Housing Shortage* in 2013, Shelter estimated that between 100,000 and 150,000 houses are required in the UK per year (Shelter 2013a), and in its 2013 report, *Homes for the Next Generation*, co-authored with the accountants KPMG, it observed that the 'consequences of this shortage are stark . . . Having a secure family home has also been linked to educational attainment, health and wellbeing' (Shelter and KPMG 2013: 3).

Affordability is measured as the ratio of household income to house price, and price has historically been 3.5 times the average household income. Duncan Bowie discusses the wide variation in affordability in the UK, and notes that the average house price in England was 7.63 times the average income in 2015 (Bowie 2017: 87–8). The most extreme levels of scarcity are in the prosperous South East and around London, as well as in cities such as Cambridge and Oxford, partly because of the lack of available land, including 'green belt' protection, and its cost. In the London region, the demolition of 'slab-block' housing schemes, such as the Aylesbury and Heygate estates in south London, has led to the replacement of social housing units with new, so-called affordable homes, which are marketed at rates significantly above the income level of the local people who previously lived there, resulting in the estate's residents having to leave the area (Wainwright 2015; Bowie 2017; Minton 2017). Lack of affordable housing also affects rural areas such as the South West and the Lake District, where local incomes are low, demand is high, and private landlords can ask for higher rents which mean that lower-income households are unable to gain secure or permanent housing (Bowie 2017).

While during the post-war period up to the mid-1960s, both Conservative and Labour governments built between 300,000 and 400,000 homes a year (Pepper 2015), a range of complex factors have led to a chronic shortage of affordable housing today. These include lack of housebuilding by successive governments, the sell-off of council housing through the 1980s 'right-to-buy' scheme, restrictions on the land that is available to build upon, and a housing market that is dominated by a few housebuilders (Bowie 2017; Pepper 2015). Since the 1980s, British people have been encouraged to invest in housing as a personal financial asset through 'buy-to-let' mortgages and low mortgage-borrowing incentives to help first-time buyers 'get on to the housing ladder' in their 20s and 30s (Ryan-Collins et al. 2017: 109–58; Bowie 2017: 36–7). The profits of housing developers and the large housebuilders are increased by 'pooling' land, the intentional restriction of the supply of houses to inflate prices and demand (Ryan-Collins et al. 2017: 197–8; Bowie 2017: 66–7), and by maximising the numbers of homes they can build on a piece of land (in 2010, two of the UK's major developers, Barratt Homes and Berkeley Homes, had profits of £2,096 million and £672 million respectively; RIBA 2011: 20). In addition, planning regulation has been weakened so that developers can manipulate their responsibility to provide the required ratios of affordable housing in new developments through 'viability assessments' which present affordable housing provision as an unprofitable burden, resulting in severe reductions in provision or even entire removal of this provision (Park 2017: 12; Bowie 2017: 114–18; Wainwright 2015). The lack of truly affordable housing is therefore one of the most detrimentally negative 'affects' in housing security and wellbeing in England at the present time.

While it is not a single solution to these highly complex political and economic housing issues, space standards, when properly used, provide very good evidence of how positive housing enables individual and societal wellbeing. Since the 1918 Tudor Walters housing report, which prescribed standards that were 'very generous for their time' (Park 2017: 54), UK housing history has featured a series of positive recommendations and policies. The Tudor Walters Committee informed the 1919 Town Planning and Housing Act, which addressed poor living conditions in towns and cities with its requirements to reduce the density of terraced housing and volume of rear elevations to improve light and air circulation, and increased size of council houses (Swenarton 1981). In the mid-twentieth century the 1942 *Beveridge Report on Social Insurance and Allied Services* and the 1944 Dudley Committee recommendations on minimum room sizes led to the 1949 Housing Act, which further improved social housing design, including elderly living needs (Swenarton 1981; The Survey of London 1994).

In 1961 the government-commissioned housing design report *Homes for Today and Tomorrow*, also called the Parker Morris space standards, laid out detailed requirements for the proportions of space to be designed into the various rooms of family homes, stating that the design of space should 'concentrate on satisfying the requirements of the families that are likely to live in them' (Parker Morris 1961: 4). The report also provided detailed design recommendations for shared space, energy efficiency, and the integration of housing into its urban setting (Drury 2008: 403–5; Carmona 2010).

Responsive to the changing demands of work and domesticity in households' living patterns, the report stated that 'The human problem for the future in the design of flats and maisonettes is to provide for people who live in them an environment which is as workable, and as satisfactory, as for people who live in houses' (Parker Morris 1961: 28). The report contains detailed tables and explanations about increased proportions of space for households, depending upon the number of people: for example, 79 sq m for a five-person flat, and 30 sq m for a one-person flat (Park 2017: 53). Internal cupboard space and external shed storage were also defined across household sizes: 1.4 sq m storage was allocated inside and outside for households of 4–6 people. For households of 1–3 people, the same external storage was supplied (e.g. a bike shed), and 0.7 sq m internal cupboard space (Park 2017: 23). Intended for all private and social housing, the recommendations became compulsory for all local authority council housing in 1967. Estates designed and built by local authorities using the Parker Morris standards included the Alexandra Road estate in Camden, Central Hill estate in Lambeth, and Robin Hood Gardens in Tower Hamlets (all in London), and Park Hill in Sheffield. Unfortunately, the Parker Morris space standards were abandoned by the Conservative government in 1981, having been deemed politically, socially, and technically unsuccessful. Financial cuts to local authority funding during the 1970s and poor design and concrete fabrication technologies combined to produce fierce public and governmental antipathy to estates that became known as 'failed' slab-block housing schemes (e.g. Carmona 2010; Drury 2008).

The value of Parker Morris as a historical basis for improving housing has, however, been revitalised over the past decade by architects and planners. This was partly in response to the 2007–08 housing crash in which large numbers of under-sized flats built during the previous 'boom' decades failed to sell, together with the increased pressure on higher quality affordable housing (Park 2017: 11). Strategies for tackling these issues of space and poor design have included the work of 'affordable housing consultants', with the Housing Association Training and Consultancy 2006 report for the Greater London Authority updating contemporary households' activity spaces for washing, eating, sleeping, socialising, and requirements for interior and exterior storage (HATC 2006). This research fed into the *London Housing Design Guide Interim Edition* (Mayor of London 2010a). Alex Ely of Mae Architects, one of its authors, has eloquently highlighted the historical significance of the Parker Morris standards and the benefits of the flexibility of spatial ratios to wellbeing (Mayor of London 2010b; Ely 2015). Also published in 2010, the Commission for Architecture and the Built Environment's *Space Standards: The Benefits* report concluded that space standards improve wellbeing by enabling sociability and privacy, aiding children's educational activities and flexible cross-generational life–work balance, reducing overcrowding and 'creating a potentially more stable housing market, driven by a more complete understanding of long-term need and utility rather than by short-term investment decisions' (cited in Park 2017: 39).

In 2015, following the government's Housing Standards Review of 2012–15, the Nationally Described Space Standard was introduced, which requires all affordable housing by local authorities to use minimum space allocations, including room height standards, that are in line with the 1961 recommendations: for example, a five-person flat should be 74 sq m plus 3 sq m built-in storage, and a one-person flat, 39 sq m with 1 sq m storage (Park 2017: 53).

Deborah Garvie, Senior Policy Advisor at Shelter, has drawn attention to the importance of the rhythms and patterns of use within a modern home, where pram and bike storage may be key indicators of what home means for a cross-generational family today (Garvie and Shelter 2015). Sarah Wigglesworth's research-council-funded project, DWELL, presents innovative, flexible housing design for the elderly through an engaged consultation process with residents in Sheffield (DWELL 2016; Park et al. 2016). Together with studies done by Shelter, RIBA, and CABE, Jeremy Till, Tatiana Schneider, and Nishat Anwan's research into historical and global approaches to 'spatial agency' (Anwan et al. 2011) also emphasises the benefits of space provision for enabling the diverse lifestyles and needs of today's households.

As I noted above, there are many complex physical, political, and financial factors that contribute to the short supply of affordable homes; also, space stand-

Figure 8.4 Holmes Road Studios, Kentish Town, London: Peter Barber Architects (photo: the author 2015)

ards are not universally agreed upon within the housing industry (Park 2017: 3–4). Urban developers such as Pocket and The Collective, a 'co-living' property company, argue that under-sized 'micro-housing' is needed for aspirational and middle-income young professionals in London. A rather different example of micro-housing is Peter Barber Architects' Holmes Road Studios for Camden Council in London, as part of his practice's specialism in working with councils and charities to provide accommodation for the homeless. This temporary housing scheme recalls the almshouse courtyard, and while the interiors are small, it provides a more dignified sheltered transition from living on the street to permanent housing than traditional hostel typologies (Peter Barber Architects 2017; Rawes and Lord 2016).

The discussion above shows that housing which is designed with the aim of supporting the different and varying needs of cross-generational and differently incomed households is critical to the equality and wellbeing of British society today. Following Spinoza's philosophical analysis of ratios of wellbeing, these examples of good housing design represent positive mental, physical, spatial, and environmental affects: for example, spatial ratios are intimately tied to enabling cross-generational households to flourish, from the young to the old. The benefit of good-quality spatial ratios and affordability are currently under severe political and economic stress, but research from built environment professionals, charities, and individuals involved in housing provision shows that there is strong historical evidence that these spatial and physical factors directly improve societal wellbeing: good housing design aids society to *live well* in its homes. Just as Spinoza highlights the ethical benefits to society that come from the *conatus*'s ability to care for itself, housing welfare is increased by good-quality, affordable housing design.

Spinoza's ratios of wellbeing therefore share a concern with housing welfare in the Dutch seventeenth-century almshouse tradition, and with twentieth-century and present-day approaches to socially directed and *really* affordable housing design. Underlying each of these subjects and historical periods is the suggestion that an *ethical* society is one which defines human needs as dissimilar, rather than imposing inflexible or unresponsive standards or universals. Instead, our capacity to live well, as individuals and as a society, is better achieved if we understand there to be dissimilar but inherent relational needs between our housing provision, especially for those who do not have strong economic powers of income or mobility, and our wellbeing.

Notes

1. There are, however, important examples of modernist social housing throughout the twentieth century which do successfully employ rational principles of spatial and geometric design that are beneficial to individual and community wellbeing, including European estates by architects such as Kramer, Oud, Sharoun, Taut, and Le Corbusier. In the UK, estates such as Bevin Court, Park Hill, Alton East and West, and Alexandra Road show how rationalist housing design approaches can produce high-quality

social housing that residents value and contemporary architectural professionals admire.
2. Goose and Looijesteijn (2012) note that research on this housing typology remains slight.
3. Goose and Looijesteijn observe that the English almshouse derives from the pre-modern institutions that provided social care for the poor, sick, and elderly, including the monastic 'farmery' or infirmary, or 'lazar houses, spitalhouses, bedehouses, Godshouses, maisondieu, hospitals'. In the Netherlands, the almshouse derives from different regional, private, or voluntary social care institutions, which might be called 'gasthuis, kameren, godskameren, godshuis, weduwenhuis, provenhuis, aalmoeshuis and hofje' (Goose and Looijesteijn 2012: 1051–2).
4. Inhabitants of almshouses were not homeless, but were from the 'respectable middle and lower middle classes' who were relieved from 'dishonourable public poverty' (Goose and Looijesteijn 2012: 1056). By the eighteenth century, the majority of Dutch almshouses catered for unmarried or widowed elderly women, or those who could not live with their families (Goose and Looijesteijn 2012: 1059).
5. One hundred and thirty-three were established in the Dutch Republic in the seventeenth century, more than double the number in the previous century, but their number dropped to sixty in the eighteenth century (Goose and Looijesteijn 2012: 1054).
6. In 2016 a privately rented, one-bedroom flat in The Hague was €963 per month (Numbeo 2017). The monthly rent for a one-bedroom almshouse in central London was £668, in contrast to the £1,480 average private monthly rent in the same area (Jones 2016).

9

The Greater Part:
How Intuition Forms Better Worlds

Stefan White

When you die this means that all the different kinds of extensive parts that make you up disappear; this means that they go off into other bodies i.e they effectuate other relations than yours . . . [I]f you had in the majority of your existence, inadequate ideas and passive affects . . . this means that what dies is comparatively the greater part of yourself . . . [S]uppose that you have succeeded in achieving mostly adequate ideas and active affects [. . .] [then] it is the other way around . . . Ethical joy is a correlate of speculative affirmation. (Deleuze 1988c: 29)

This chapter sketches out an architectural response to Gilles Deleuze's 'expressionist' re-examination of Spinoza's *Ethics*. Deleuze argues that Spinoza's 'expressionism' creates the potential to positively reframe a repressive Platonism which had been carried through the Enlightenment within Cartesianism (Deleuze 1992: 322; 1994: ch. 4). Deleuze explains that predominant – 'representational' – systems of thought privilege processes of identification on one or other side of an essentialist, Cartesian dualism – acting either as empiricism, 'referring to causality within Being' (emphasising functions) or as idealism, 'referring to representation in Ideas' (emphasising forms) (1992: 333–5). Spinoza is a key influence for Deleuze's contribution to what he sees as the 'task of philosophy itself', defined as the need to 'overthrow Platonism' by 'abolishing the world of essences *and* the world of appearances' (2003: 253). He claims that a triadic Spinozist principle of *expression* simultaneously addresses both of these essentialist tendencies in traditional epistemologies, producing an approach more able to account for the power or potential of creative acts. Deleuze's interpretation of Spinoza holds that his triadic epistemology creates an ethological philosophy for life through the 'curious . . . intervention of a type of relative proportion' (2003: 15). To explore these ideas in an architectural context I will discuss correlations between Deleuze's critique of 'representation' and the ideas of the architectural theorist Robin Evans, using Evans's concept of 'projection' as a 'non-representational' way of thinking through the role of architectural practice, processes, and products.

Evans examines the design processes of architects, exploring the relationship between the kinds of drawings and models they make and the buildings and spaces that 'result'. I will outline parallels between Deleuze's critique of Descartes and Evans's assessment of the inadequacies of the predominant architectural epistemology – 'picture theory' – to set out the terms for arguing that Evans's *projective* response to the reductive translation processes of 'picture theory' produces an expressionist alternative to representational practices in architecture (Evans 1995: 358). Evans's account of architectural production creates an embodied triadic structure by reiterating the importance of a normally repressed third term, which he describes in terms of 'projection' and which I explore in terms of the triad of project-process-product as a way both to ground the excesses and to transcend the limits of traditional – dualist – design epistemologies. This triad is used to explore how the third kind of intuitive knowledge of 'intensive essences' enables an affirmative production of ethical joy in architectural practices, seeking to make 'adequate ideas and active affects' the greater part of their expressive capabilities (Deleuze 2003: 15).

The chapter comprises three sections, responding to the triad of expression of architecture – Project / Process / Product. In the first section, 'Project', I will describe three key features of the Deleuzian reading of Spinoza: the critique of inadequate or representational ideas, the process of distinguishing more adequate ideas, and the nature of active affects or intuition. In the second section, 'Process', I will describe how these features might correlate to the practice of architecture by exploring Evans's critique of existing representational theories of architecture and explaining his projective alternative. In the third section, 'Product', I will summarise the potential ethical and aesthetic implications for an expressionist account of architecture.

Project

Deleuze argues that Spinoza provides a radical (but subtle) response to the inadequacies of Cartesianism. For Spinoza, it is absurd for Descartes to maintain that *'there are numerical distinctions that are at the same time real or substantial'*, which for Deleuze is the same as saying, 'there exist Substances sharing the same Attribute' (Deleuze 1992: 28). If substances can share attributes, then attributes would be distinguished only by their modes, but modes are already *modifications of* substance (Deleuze 1992: 30). If two substances shared an attribute (if two powers shared the same qualities), then they could only be *numerically* distinct and there would be no *real* distinction between them (Deleuze 1992: 31). For Deleuze, the opening of the *Ethics* reveals the consequences of this error in Descartes, an error that results in substances 'being reduced to the mere possibility of existence and attributes mere indications of such possible experience' (1992: 36). In contrast to their indicative status for Descartes, Spinoza conceives attributes to be really, substantially distinct.

In Deleuze's reading, the distinction of attributes is the qualitative composition of an ontologically single substance. For substance to have infinite attributes but be singular, attributes must be understood as the distinctions that compose substance, and these distinctions take the form of its component qualities. Rather than named qualities, these real distinctions are 'purely qualitative, quidditative or formal', excluding any division (1992: 182). These real distinctions are therefore productive of substance and serve to compose it. It is in this way that substance cannot be merely possible existence nor attributes mere indications. The distinctive process of attribution is never numerical nor modal, but of real power or potential. For Deleuze, the insistence on this reality of distinction between attributes enables a fundamentally positive conception of difference (1992: 38).

Deleuze claims that Spinoza must be understood as attempting to produce such a positive account of difference, or else his definition of substance may be seen to determine modes of existence without those modes of existence appearing to have any effect on substance itself. For example, he argues that the concept of expression has to explain how 'the many' constitute 'the one' and 'the one' produces 'the many', because if 'what causes' is a substance *indifferent* to the modes of being that it 'causes', it is merely Cartesian essentialism by another route. Here Deleuze seeks to remove any 'indifference' in the process of attribution by giving difference its own concept, so that we are able to think beyond the Cartesian conception of difference as a negative comparison to identities.

To do so, he begins with difference as a prior relation of production, understood as 'Difference-in-itself'. Deleuze assumes that the elements of Spinoza's ontology – substance, attributes, and modes – are not fixed or static objects but are explicated as if they were emerging from this 'prior relation', and here the concept of expression is understood as a relational logic which is also generative (Durie 2002a). From this point of view, the qualities or attributes of nature are themselves the distinctions which constitute substance and subsequent distinctions are explications of that process. 'The Attributes are, according to Spinoza, forms of being which do not change their nature in changing their "subject" – that is, when predicated of infinite beings and finite beings, Substance and Modes, Gods and creatures' (Deleuze 1992: 49). Deleuze argues that Spinoza makes clear in E IP25C the distinctive, explicative role of expression: 'Particular things are nothing but the affections of the Attributes of God; that is Modes wherein the Attributes of God are expressed in a certain and determinate way.' For Deleuze, this is the decisive strategy of expressionism, since it makes real distinctions 'capable of expressing difference within Being' and brings about the 'restructuring of other distinctions' (1992: 39).

While the restructuring of distinctions begins with the idea of a prior relation of expression being central to the process of distinguishing between a nominal and a 'real' distinction, there is much more to do to make this approach coherent. Deleuze employs 'a theory of distinctions' (1992: 37) developed from the work of Henri Bergson to argue that the relation between identities or the names or numbers of things produces two types of numerical, quantitative distinction (what Deleuze calls 'differences in degree' and 'degrees of difference') as well as differences between attributes, powers, or qualities (Deleuze follows Bergson to call them 'differences in kind'; 1988a: 43; see also Durie 2002b). In this approach, all three of these distinctions are *real* – either abstractly in terms of potential (virtual) or specifically in terms of concrete things (actual) – but never merely generally possible or imaginary. It is this triadic understanding of distinctions which underpins Deleuze's reading of Spinoza's triad of expression as a serial and parallel relational structure which has no *indifference* between its substantial character and its modes of existence.

The process of distinction or differentiation is understood by Deleuze in terms of the concept of expression, which he describes as a differential attribution. In Deleuze's reading, this restructuring of distinctions produces a triad rather than a dualism, because each pair of terms has a third, intervening one – the affected or affecting 'body' in which the subject or object is, in fact, expressed for *real* as 'sense'. The expression of sense is the substance of the composition of the bodies that make up the concrete real, actual specific *and* the real, abstract virtual world. An expression is always a double attribution or articulation – in what is expressed *and* what it is expressive of. Expression here appears as an attributive double relation between form and absolute: 'Immanence is the very vertigo of philosophy and is inseparable from the concept of expression (from the double immanence of expression in what expresses itself and of what is expressed in its expression)' (Deleuze 1992: 180). It is this double or simultaneous articulation or explanation of the existence of the expression through the expression of existence that makes the theory 'immanent', because God does not cause this world from the outside but is part of it, in the sense of a pre-existing relation, by way of the pre-existing relation of expression.

However, in order to *make* sense, this double articulation has to be very particularly understood. If it is clear that all modes are expressions of substance, expressions of the power of existence, then the power expressed by the attributes must be of a different kind to that expressed by the modes, otherwise they would remain indistinct. So for Deleuze, if attributes are dynamic qualities whose essences correspond to the power of substance, then modes may be thought of as a quantitative differentiation of those qualities: 'God's power expresses or explicates itself modally and only in and through such quantitative differentiation' (1992: 183). However, the power of substance is determined through its expression as an existent mode – through the attribution of the qualities of substance in that existence: 'The production of Modes, it is true, takes place through differentiation. But differentiation is in this case purely quantitative. If real distinction is never numerical, numerical distinction is conversely, essentially modal' (1992: 183).

While this appears to relegate the status of modes, Deleuze's extrapolation of Spinoza's concept of expression 'asserts immanence as a principle and frees expression from any subordination to emanative or exemplary causality' (1992: 180). Expressions are at the centre of the continual and iterative *double* process of attribution that generates modes *and* in turn modifies substance. In this process of distinction between attributes and modes, what is expressed by the modes is a 'modification' – *a particular* change in the *power of existing* of the body expressing its power of existence, whereas what is expressed in the attribute is *the power of existing itself*. This distinction between attribute and mode is doubled by the simultaneous distinction between attribute and substance. Attribution differentiates simultaneously *qualitatively* and *quantitatively*, producing modes through a quantitative differentiation and constituting substance through a qualitative differentiation.

In Deleuze's approach, attribution is a process of distinguishing or 'differentiating' substance (it makes substance *differential*) and substance is a 'qualitative multiplicity', an infinity of attributes which have qualities but not quantity (Deleuze 1988a: 42). They have no quantity in the sense that they cannot be counted or divided without changing their nature. However, there is a second aspect of attribution that is also a differentiation. That is to say, the existence of modes expresses an essence of substance, as a modification of the power of acting, by way of instituting a differentiation of degree of an attribute or 'kind' of substance. Hence there may be an infinite number of modifications of a single attribute and the essence of the mode expresses degrees of power of substance as ('intensive') *degrees of difference*, but the actual existence of the mode simultaneously expresses ('extensive') *differences in degree*.

> The individuation of the infinite does not proceed in Spinoza from genus to species or individual, from general to particular; it proceeds from an infinite quality to a corresponding quantity, which divides into irreducible intrinsic or intensive parts. (Deleuze 1992: 199)

The division into intensive parts is the first stage of the distinction, where essence becomes the potential of an actual existence by first becoming a divisible relation. For Deleuze, these intensive differences or *degrees of difference* are, for example, the differences between different 'whites'; that is to say, white is seen as an attribute of substance and white*ness* an intensive differential, intrinsic to the 'attribute' white. The existence of a particular white, the existence of a particular mode of whiteness, would therefore have precise *degrees of difference* from another particular white (Deleuze 1992: 196). These *differences in degree* and *degrees of difference* are *expressive* of the *differences in kind* which are attributions of substance. 'Substance expresses itself, attributes are expressions, and essence is expressed . . . we confuse substance and attribute, attribute and essence, essence and substance as long as we fail to take into account a third term linking each pair' – their expression (Deleuze 1992: 27). Modes are expressive of the differences in power which constitute them and which, in turn, they contribute to. The *expression* that produces a specific mode is a particular degree of the differences of relation with its *expresser* – a qualitative differentiation or attribution of substance. What is *expressed* is only what actually comes to exist through the expression – a difference of relation with the outside of the expression. Substance is pure quality that is distinguished through its attribution as particular qualities in the production of modes which have certain *quantities of quality* and which also have a *specific quantity of those amounts of quality*. Actual modes of existence are here understood as a quantitative distinction, but one which is doubly articulated as an intrinsic *and* an extrinsic mode.

In this formulation, modes are a particular degree of power (of substance), and a particular modification of that power (of substance). The first form of modal existence is spatial, intrinsic, or kinetic ('latitude') and can be understood as 'the expression itself'. The other is durational, extrinsic, or dynamic ('longitude') and can be understood as 'what is expressed'. Deleuze argues that these relations of longitude and latitude construct a map of a body, and the plan or 'plane', which these maps each partially describe, constitutes nature or substance or 'the plane of immanence' (1988c: 122–7). In this reading of Spinoza, the 'relations of movement and rest' – the kinetic aspects of a body – are the enduring relations that characterise it and form its physical 'extension'. In addition, 'the dynamic relations of capacity' that a body has at any moment, by virtue of its particular arrangement of extensive parts and the relations that compose them, are an 'intensive' aspect of the 'body'.

> All modes participate in the power of God: just as our body participates in the power of existing our soul participates in the power of thinking. All modes are also parts, a part of the power of God, a part of Nature. (Deleuze 1992: 146)

Deleuze argues that Spinoza's approach excludes causality between ideas and things, between thought and extension, and this sets him apart from the ancient tradition (Deleuze 1978: lecture of 24 January). In his reading, 'Substance is . . . the power of existing in all forms, *and* of thinking in all forms' (Deleuze 1992: 198). This immanent definition of power means that a distinction of existence is also a distinction of knowledge (with each understood as different *kinds* of affective powers). It is only subsequently that knowledge is considered in terms of separate attributes – relations of extension or thought. A 'body' is defined both by the affect it may have on others and how it may be affected – and these affections may be simultaneously mental and physical.

It is crucial to Deleuze's restructuring of the hierarchy of distinctions of Cartesianism that the two forms of a body – the intrinsic and extensive modes of existence – not be once again conflated to either thought or extension. A key capability of the expressionist triad is therefore how it complicates and explicates the mind–body dualism of Descartes, because expressions can be both mental and physical at the same time (in parallel). The two forms of modal existence are able to express the attributes of *both* thought *and* extension to different degrees in different contexts. For Deleuze, this move produces the possibility of non-identity with causes, removing a causal chain to replace it with his expressive logic. Instead of the expressions of the world being interpreted as either an object or an idea, here *both* object *and* idea, in parallel, have their own duality.

Ultimately, Deleuze's diagram of expressionism is the insistence that reality has three components not two; that the differential power of the expresser always precedes the formation or functioning of any aesthetic or ethic. Beginning with that which *expresses* is intended to avoid a reductive identification of the expression with what is expressed and any ensuing internal conflict of prioritisation. The dualism of representation allows this identification to be performed from either one of two external and reductive perspectives: one outside the situation in time (what is it in essence?) and the other outside the situation in space (what is it in appearance?) Rather than taking a complex phenomenon and removing either time or space in order to reduce the problem to identification of instances or states, Deleuze argues that we should instead articulate it intensively and extensively, in parallel. When we look at any expressive object, person, drawing, or problem from this non-representational perspective, we must always ask the simultaneous questions: of what is it expressive *and* what does it express?

According to Deleuze, when Spinoza begins the *Ethics* by defining God before proceeding to define human reason, this hierarchy creates a distinction between kinds of affection (Deleuze 1992: 218). God is considered affected in an infinite number of ways, and absolutely unlimited. God is the cause of everything, the cause of all his affections, and so cannot suffer them. God is wholly adequate because he is the cause of all his own affections – he is *active*. On the other hand, therefore, affection not explained by the nature of the affected body must be explained by the actions or influences of other bodies – it is reactive or *passive*. Existing modes do not naturally meet this requirement, because they do not exist by virtue of their own nature but are expressions of relations with other bodies – 'compositions of extensive parts that are determined and affected from the outside, *ad infinitum*' (Deleuze 1992: 219). This distinction means that the affections of existing modes are defined at the outset as, and tending to remain, passions. God or nature here is the potential of existence, and our 'soul' (the expression of our being which is not our body) is seen as the proportion of the power of existence that we have the potential to possess or express, since at the same time that we participate in nature we also contribute to its composition.

The capacity to act, or power to be affected, is a constantly varying dynamic – the 'melodic line of continuous variation of affect' (Deleuze 1978: lecture of 24 January). In these terms, we never cease to pass from one degree of 'perfection' to another, however minuscule the difference between these states. Here the proportional variation of our state of being in relation to the affect of the world upon us is produced through our difference from or correlation with it – how it changes us, how our idea of the relationship is changed and our will towards it found agreeable or disagreeable. Deleuze describes the process of developing the capability of affecting and being affected in terms of three ways of living, three kinds or stages of expressions of being, which correspond to three kinds of knowledge. First, the knowledge we hold of external bodies through the affects they have upon us (signs); second, the knowledge we gain of our own body through these affects (reason); and third, the knowledge we gain through the understanding that we have of ourselves and external bodies (intuition) (Deleuze 2003).

All bodies are striving for or desiring of agreeable compositions and the minimisation of disagreeable ones – from combinations and reactions between chemicals to individual human beings and social systems. Agreeable compositions (joys) increase our ability to act, whereas disagreeable ones (sadnesses) reduce our capabilities. The seeking of agreeable compositions is induced by it resulting in an increased capability to act – it is joyous. Such is the nature of composition and duration that a body that did not seek agreeable compositions, that did not strive, would cease to exist or would not have duration enough to be considered (affecting other bodies to that extent) as an individual. What drives this striving, as it were, is in fact the potential of entering into agreeable compositions, of augmenting duration, in terms of the actual possibilities which arise to do so, not in terms of an external notion of mortality.

The process of developing as an individual begins from a nascent state of encounter with signs of existence. This first state of knowledge then leads to the development of the capability of reason, since the relationship between imagination and the world of effects may produce knowledge of the causes of those effects through the correlations of experience. The second state of knowledge produces forms of reason through experience and associations via the imagination, enabling repeatable encounters. These enable development of a third kind of knowledge whereby the relations between the causes which enable repeatable encounters may intuitively be engaged with – which is to say, actively selected.

In Deleuze's expressionist system, it is joy that drives the production of life and knowledge, because we must have a positive, joyful conception in order to form a reasoned common notion, which must compose or balance both negative and positive aspects in order to overcome the passive understanding of effect. Without the desire to determine a more positive outcome we would not construct a reason by which we could do so. In turn, the effort of creating intuitive actions is a pure expression of the joy of speculation as the active pathway to an increase in capacity. The expressionist notion of embodied non-representational knowledge acts towards an increase of capacity through striving towards the outside, because the only way to produce and increase sense is through open affective relations with external bodies. The triad of expression is a way of articulating the kinds of knowledge that a body may gain and hold. In this approach, the disciplining, categorising nature of dualist knowledge production is insidiously reversed, so that the forms or functions of an object are no longer thought in terms of identifications with good or error, but in terms of their contribution to or participation in the production of qualitative differences between modes of existence.

Deleuze argues that this is where Spinoza poses the central question of his ethical analysis. How might we ensure that our passions take up the smallest part of us? How might we be free? Deleuze describes the sense of joy as 'the truly ethical sense' because 'Spinoza's naturalism is defined by speculative affirmation in his theory of substance and by practical joy in his conception of modes' (1992: 272). It is in this sense, therefore, that the formation of embodied intuitive knowledge is at once a creative, expressive undertaking *and* an ethical practice. In this understanding there is no advantage in asserting the existence of one beautiful form over all others, but striving to make more agreeable ones, in particular times and contexts for particular purposes, is an ethically joyful, speculative practice. Creative activity towards the outside of the subject is an ethical affirmation because it is an affirmation of the qualitative difference between modes of existence, rather than an imaginary categorisation against a transcendent system of judgement. It is a productive, real, immanent (abstract and specific) distinction, rather than a merely nominal or representational one. The opening of the body to be affected by the outside is an immanent ethical negotiation with the world, which is speculative in the sense that the outcome of this encounter is not pre-determined but simultaneously affirming of the reality of the potential produced through external encounters. It is an active, joyful affection of the body in the sense that it is an increase in the body's capacity.

Process

> The picture theory has been difficult to see through because pictures can stand for a large part of vision very convincingly ... imagination and perception are shown as pictures, because that is how they are normally described. They are not pictures, but the very fact that they are thought of in that way is very significant. (Evans 1995: 358, 370)

In Robin Evans's posthumous work, *The Projective Cast*, he tests the proposition that 'Architecture is more than the sum of its representations' (Evans 1995: ix). Through a series of detailed case studies, he makes a strong argument that architecture's creative processes cannot be adequately accounted for using traditional approaches, which he calls 'picture theory'. Picture theory is defined as a way of thinking of the process of drawing and designing that understands imagination and perception as objective representations – simple pictures of reality – using the analogy of 'translation' to correlate between future 'realities' (the buildings our designs result in) and the 'pictures' we make of them (architectural drawings). According to Evans, 'the translation analogy' has come to be the most common mechanism for understanding the relationship between drawings and buildings in the architectural process, both among the general public and architects themselves.

> [T]he assumption (of the translation analogy) that there is a uniform space through which meaning might glide is more than a naive delusion ... Only by assuming its pure and unconditional existence in the first place can any precise knowledge of the pattern of deviations from this imaginary condition be gained ... [S]omething similar occurs in architecture between the drawing and building. (Evans 1997: 153)

Deleuze argues that for each of Descartes' four faculties – of conception, perception, judgement, and imagination – the presumption of a universal subject or 'uniform space across which meaning glides' produces a different form of illusory identity relation understood as conceptions of the identical, perceptions of similarity, judgements of analogy, and finally imaginations of opposition (Deleuze 1994: 262–70). The presumption of the 'pure and unconditional existence' of these identity relations makes them the imaginary condition against which judgements of 'deviation' are made. In architecture, these four categories are related to an increasing complexity of relation between drawn image and projected object, or between imagined object and drawn expression of that object, beginning with what Evans calls the expectation of 'close alignment' (1997: 153). Evans raises the presence of these illusions of representation in architecture as more than 'naive delusions' because of the advantage they confer. He argues that they are 'enabling fictions', which serve to maintain the dualism, while they remain implicit:

> such an enabling fiction . . . has not been [made explicit] in architecture [and] because of this inexplicitness . . . on the one hand, the drawing might be vastly overvalued, on the other, the properties of drawing – its particular powers in relation to its putative subject, the building, are hardly recognised at all. (Evans 1997: 154)

Evans argues that the advantages of the easy narrative of 'translations' have ensured the architect's complicity in two 'demonstrably false' assertions operating in architecture. The first assumes that there is a direct correlation between 'idea' and 'thing', the second that the individual imagination is the site of creativity. These fictions, then, may on the one hand enable the profession to offer a determined and quantifiable service through a deliberate overstatement of the drawn object's ability to determine the built object, but on the other hand they greatly limit the embodied understanding of the creative relation between the drawn object and its subjects.

Deleuze notes that representational epistemology readily assists this delusional behaviour by being flexible enough to recognise a 'nobility' in the indeterminate processes of experimentation and learning. However, he argues that this ultimately takes the form of an homage to the 'empirical conditions of knowledge' whose traces of indistinctness or complexity are seen only as 'preparatory movements' and are all expected to 'disappear in the result' (Deleuze 1994: 166). The complexities of the creative moments in the design process are generously tolerated, but only briefly, and only until the process meets its ultimate expectation of delivering a precise translation between drawn conception and physical construction. The differences between the drawing and the subject of its translation or between the imagination and the drawings it produces are then 'good-naturedly ignored' by the common sense of the faculties, to either justify the exchangeable professionalism of the drawing or the elite expertise of the designer.

The dualism is seen either to limit the capabilities of the drawing and its product through the expectation that there is a causal relationship between one and the other, or to limit the capabilities of the drawing and its producer through the presumption that the process of identification is a purely inherent or internal one. The first posits a notion of the drawing as a pure appearance, ultimately relegating all creative interrelational aspects of the process. The second prioritises the creative process but makes it into an essence of expertise, in this case relegating the actual expressive mechanisms of transmission of cause (construction processes that produce the built form) to the realm of 'intelligible essence'. The functioning of the representational system of thought requires the institution of a common sense, and people of good sense to found it. The deliberate and malign moralism at the heart of representational thinking is, in some ways, an unfortunate side-effect of its efficacy. However, when a drawing is expected to determine the built outcome as a direct, absolutely determined codification (as a simple translation), then the inevitable differences between drawing and built resolution that in fact constitute its creative mechanism are seen only as error or lack. The differential processes which are the source of both the drawing and the architect's power are conveniently ignored through a continual insistence on judgements of identity, because shameful errors are best kept quiet about – or used as evidence of the poor comparison between reality and the possibilities of genius – which then further stokes the cycle of delusion.

Evans's epistemological project is to properly account for the limited powers that the drawing has in processes of determination, and at the same time fully recognise the understated powers it has in terms of creative production. While Evans finds the drawing overvalued in terms of its ability to have a determined or causal relation to its object, he also argues that the drawing has a power that remains unrecognised, a power that he locates in its distinctness and unlikeness to the thing it is used to 'represent'. Here, I believe, Evans may be taken to be both articulating the inadequacies of a representational dualism and identifying the positive – 'affective' – potential in the 'transitions between objects' (Evans 1995: 367). The relationship of non-identity or dissimilarity between imagination and perception in Evans's 'projective' account is directly concordant with Deleuze's critique of representational thought and the notion of non-representational thought as 'Affect'. Evans does not believe that the drawing is a code of similarity or analogy allowing translations between universal surfaces of representation; rather, he insists that the drawing 'always interacts with what it represents' (1997: 199). Deleuze argues that while the illusions of representation see difference as being a terrible lack (of similarity necessary for a perfect translation), creating and enforced by the fear of non-sense, it is instead the paradox of inherent dissimilarity in 'transitional' relationships of affect that brings potentialities into existence.

Alongside a philosophical impetus to address the character of the underlying epistemology of architectural theory, Evans builds a strategic view of the relationships between drawing, design, and construction processes in order to delineate a more adequate alternative to translation that does not see an identity relation between imagination and perception but rather prioritises the projective – differential – nature of this relationship. He picks out moments in the processes of translation between drawings and their buildings where the analogies of representation no longer hold and where the importance of the concrete interactions between the architect and the design objects are irrefutable; it is these relationships that he labels 'projective'. Evans argues that the traditional, inadequate account of all of these projective moments in the design process contains the essential assumption of representational (picture) thought – that a mental image held in the retina or the brain is of the same nature as a formal image held on paper, and conversely that these are of the same nature as the image of reality perceived by the eye or the camera. He identifies ten points in the drawing production process, which characterise moments when the translation analogy and these relationships of identity are normally operative (1995: 365–7).

These moments, which are normally described using the concept of 'translation', must, Evans argues, always involve instead a process of 'projection'. He demonstrates the inadequacy of the traditional materials of translation for determining the translated object without additional creative intervention across a range of contemporary and historical case studies. He shows that every circumstance of assumed translation also requires a creative (imaginative *and* perceptive) act to 'fill the gap' (1995: 363) between a partially adequate, objective codification of 'abstract mathematics' and a partially adequate, subjective process of de-codification of 'palpable experience' (1995: 366). Evans describes these ten types of relationship in detail in the final pages of *The Projective Cast*.

The first, simplest case is the projective relationship *within* drawings. Here, a mathematical process of projection can be employed to create one, codified, three-dimensional drawing from a series of interrelated, two-dimensional drawings (for example, a perspective from a set of orthographic plans, sections, and elevations). Here, there is a codified system for the expectation that one drawing would produce the other in a precisely determined fashion. However, the reality is that each plan and section is a creative reduction of an imagined whole and when used to create another drawing (like a perspective), further processes of reductive and expansive imagination occur. This relation is further complicated when a real object – a building for example – is put into the sequence. The complexity of drawing an object or making the object from a set of orthographic drawings (or perspectives) once again involves a creative – reductive and expansive – act of imagination *and* (more obviously this time) perception, which have to have the gaps between one and the other 'filled in'.

While Evans focuses on an example of a projective relationship *within* a drawing, for simplicity much of his historical analysis discusses how different drawing and modelling techniques involve different forms of projection and evolve together with different kinds of design products or buildings (Evans 1995: ch. 3 and 4). So, while Evans prefers 'projection' to translation because he demonstrates that creative drawing processes literally 'project' a particular set of contents into a new – expressive – form, he also applies it less literally (for example, when a sketch model is 'drawn' into an axonometric) *and* more broadly, to explain how design techniques and abilities progressively and differentially inform each other. Evans's 'projection' is an intuitive and embodied capability of the architect that enables him or her to actively iterate compositions from the content of one formal product (a drawing or model) to a new one through a particular drawing or modelling process over time or in series.

Evans argues that drawings are powerful determinates of forms through the representations they identify and fix, but are more importantly to be understood as generators of form through the emergent trajectories that they open as the expressive products of their creators and operators. He argues that it is the 'mobile' transitions between states or pictures that must be understood as the key power of creative processes, because it is here that the imagination is not identical with the idea.

Evans's projective relationships occur not only in the dynamic, embodied sense of operating *between* static representations, but also (simultaneously, contiguously) through the serial, iterative process of expression of the imagination between bodies in or across *time*. This argument involves accepting that the imagination is not a personal generator of form or ideas, with each individual genius expressing the essence of beauty, but rather a serially interactive construction of relations between expressions as drawings; between what is expressed in drawings and what techniques are able to express; between techniques and attempts to use and understand them to produce particular kinds of expression; and between these expressive technologies and the social contexts that desire or inhibit their development and dissemination.

When Evans insists that the transitions between building and drawing no longer be elided by the reductive conceptual identity of 'translation', he insists on a triadic, expressionist notion of architecture. He locates the architect's creative power not in a transcendent, autonomous imagination but in the concrete process of production of form that occurs in the active engagement between architectural bodies. It is in these relationships, which are external to but immanent (contiguous and simultaneous) with the mental representations of authorship, that the power of the imagination produces forms. Evans's relationships of projection reconfigure the traditional representational identification that removes the intervening subject. He instead presents the situation from the perspective of an embodied subject where imagination and perception are dimensions or attributes of a – connected – two-form expressive process.

This mode of analysis sees buildings, drawings, and architects as bodies that are both constituted by the relations in which they are situated and determinants of those situations. Through making the separation between imagination and perception explicit, it posits an architect as having attributes of both embodied engagement and intellectual abstraction that may be simultaneously and independently expressed. Consequently, each of his or her projective moments outside of imagination or perception can be described in relation to either the imagination or the perception. In traditional epistemologies, the different emphases that these perspectives produce may be seen to result in the dual, divergent, essentialist explanations of these relationships – as either artistic genius or professional codification. Evans's projection insists that the committed human labour of design is made an explicit part of the epistemological explanation of process and product *in parallel*, with both understood as *serially composed* manifestations of the power and potential of creative, differential relations. This concept of projection has the potential to value the architectural design process as a serial production of *embodied* differences between forms of expression *and* their contents *over time* – rather than it being understood representationally as an instantaneous and seamless translation, with all errors removed. In this non-representational paradigm, the creative – projective – act always involves an active, intuitive engagement with the outside of the existing characteristic relations of the architectural body.

This concept of projection has the potential to value architectural design as a serially projective or expressive process between *expressions* (the process or content, for example a drawing or multiple drawings in series) *and* what is *expressed* (the product or form, for example what is understood by the drawing and subsequently actually built). The expressions of an expressive process – the drawings produced by the intuitive capability of an architect – have their actual existence as a power produced in the differences between what the architect imagines and what they perceive in relation to it. Similarly, when the drawing becomes a product it is embodied in forms external to the architect or the architectural process, ultimately being built through a set of interpretations and realisations, which will all involve embodied 'projections' of those involved. While Evans demonstrates through numerous case studies that it is fundamentally impossible to reduce this process to a direct translation, he also shows that it is *undesirable*, precisely because these embodied moments of projection are how creative acts are stimulated.

The key implication of Evans's project is that the political and ethical content of his account of architectural processes – that which makes it critical in architectural discourse – is that every formal aspect is understood as being produced not through a solipsistic autonomy, but through a fundamental encounter with the outside of the subject. Evans's insistence on an account of the relationship between drawing and architect and building, as a projection into the future that occurs both in an architectural body and a body of architecture, is a plausible process by which we may argue for an account of architecture that does not overvalue the author but instead insists on the serial, historical, and external constitution of architects and their discipline. Valuing the embodied relations that enable the movement from one form of content to another via the expressive affect of the drawings, models, and buildings of the design process produces a positive role for the architect as the explicit relation between discipline and society, rather than, say, having to resist social pressure to protect a creative 'autonomy'. It is such compositions of difference (between drawings, disciplines, cultures, and individuals) that are themselves composed into *projects*, which exist through and within architecture but always beyond it. Evans seeks an adequate explanation of architectural composition that does not suffer the naïve delusions of dualist thought but the embodied, differential, expressionist account of Deleuze precipitates a more adequate explanation of the composition of *architecture itself*.

Product

This Deleuzian shift to a third – projective – kind of knowledge generated through action or capability engages with the composition of architecture as a discipline, not just the compositions of the discipline of architecture. Evans saw that the power of the drawing is produced in the differential relationships between the architect and the production process, and between the drawing and the process of construction. He understood the critical creative impulse as an explicit opening of the architects' compositions to the outside of their habit. However, such compositions are conventionally restricted to formal expressions from within the discipline and its 'culture'. The predominant 'diagrams' of 'professional' and 'critical' creative practice require that we defend an authorial position through restricting the extent of our 'criticality' (also understood as a breaking of norms or habits) as relating to the historical constitution of the discipline and/or the interior processes of architectural design (see Ghirado 1994; Eisenman 1995; and White 2014). The habitual architectural body defines its boundaries by setting limits on its ability to engage beyond its existing characteristic relations. Unfortunately, this is an increasingly ineffective attempt to ensure disciplinary survival in the face of commercial pressures that only serves to limit our powers of creation and our capacity to evolve new selves, practices, and societies.

The aim of this account is to show that the architectural body can no longer exist within the categories of 'architect' or 'architectural drawing' nor the categories 'architectural discipline' or 'architecture', because the creative impulse of the individual architect is at once determined by the discipline *and* serves as its constitution. It is the active compositions of the architect that constitute an architectural *project* (through their agency and movement), and in turn their and others' multitudinous architectural projects that actively constitute 'the discipline' within and through society. The particular expressions of architecture's products or processes – any of its individual drawings or representations, including the actual building – are now also to be valued in terms of the project they constitute rather than only the product they delineate. In turn, multitudinous architectural projects now actively constitute 'the discipline' within and through society.

The architectural body defines its boundaries by setting limits on its ability to engage beyond its existing characteristic relations, but in so doing it limits its power of creation, its capacity to act and to compose joyfully. An adequate architectural account makes *explicit* disciplinary limits produced not only by relations of formal expression, but also the relations of content of those expressions, and in so doing expands them. In this approach, the 'architect' is just a particular case of the citizen. Citizens are understood as embodied subjects distinguished and produced through changes in characteristic relations and affective capacities. Cities are understood as bodies described by a set of social and physical capabilities, constituted by the expressive forms and content of their citizens in relation and over time in the formula Cities = Architecture + Citizens. Creative and active engagements are the production of real distinctions of affect between bodies, necessarily external to the subject, and necessarily distinctions in both expression and content. Questions normally excluded from architectural discourse begin to re-enter as a personal *and* structural imperative. How does architecture contribute to social progress? How can architecture create inclusive environments? What difference can the actions of an architect, *this* architect, make? Here, architecture must actively differentiate itself from within society rather than create, presume, or pretend an exclusive identity separate from it. Extending and sharing the production of capabilities and powers of expression beyond the 'architect' and 'architecture' constructs the discipline as a composition that minimises its production of inadequate ideas and passive affects. Accounting for and understanding how to ensure that adequate ideas and active affects are comparatively the greater part of our individual and collective architectural selves may be a joyful and ethical activity consisting of positive affective speculations outside of our habitual identifications.

10

Slownesses and Speeds, Latitudes and Longitudes: In the Vicinity of Beatitude

Hélène Frichot

Practising Immanence, a Life . . .

How does a reading of Spinoza's *Ethics*, further underwritten by the conceptual diagrams of the philosopher Gilles Deleuze, suggest ways of following the ethico-aesthetic acts of a creative practitioner working at the threshold between art and architecture? In what follows I will introduce Deleuze's reading of Spinoza's *Ethics*, specifically in relation to what he calls the 'three ethics' (Deleuze 1998) and 'the three kinds of knowledge' (Deleuze 2003), understood as series of thresholds through which a mode of life strives in order to arrive at a summit where experimentation gives way momentarily to a calm refuge called beatitude. Beatitude is a concept, or rather a state of blessedness, to which Deleuze dedicates a chapter in his book, *Expressionism in Philosophy: Spinoza* (1990). He returns to this concept in an essay he wrote shortly before his death, 'Immanence: A Life. . .' (2001), where, alongside a consideration of beatitude, he meditates on 'pure immanence' and 'a life'. Beatitude is a concept that can be easily overlooked if one refers only to the English translation of Deleuze's late essay, where it has been sadly diluted, or quite simply mistranslated, as 'bliss', which is another passive-affective modification altogether. It is Deleuze's long engagement with Spinoza, which spans from the onset to the conclusion of his academic life, that explains the presence of this concept of beatitude, and confirms the central importance of this thinker for Deleuze. And so, after Spinoza, Deleuze tells us that beatitude is a mode of life in which one achieves the maximum of active power or force of existing, and the minimum of reactive passions.

For the most part, along the way, a mode of life struggles and strives, and as Deleuze explains, those who experience only inadequate ideas, at the first level of knowledge, while they should be judged as no lesser beings, remain 'ignorant of causes and natures, reduced to the consciousness of events, condemned to undergo effects, they are slaves of everything, anxious and unhappy, in proportion to their imperfection' (Deleuze 1988c: 19). Our power of acting is increased, Deleuze explains, 'proportionately' (1988c: 28), and something of a 'transmutation' takes us across the threshold from one level to the next of Spinoza's three ethics; the three levels of knowledge. If we have experienced a proportionally greater number of affections, and if these have been converted into adequate ideas, then we approach the greatest proportion of active (and not passive) affects, and thereby experience joy. The greatest achievement in such a formula is beatitude, which designates the greatest proportion of adequate affects, where extensive existence and intensive essence momentarily align. The question of proportion takes hold in the inflations and deflations, the expansions and contractions of expressions of modes of life and their respective capacities in relation to their immediate environment-worlds. When the mind becomes a cause of its own ideas, and the body that of its actions in relation to an infinite milieu or plane of pure immanence, this is where the greatest proportion of active affects is achieved. This situation, which may well be short-lived and fleeting, places a mode of life in the most intimate relation to God-Nature, or absolutely infinite substance, and in the vicinity of beatitude (E IP13). Following Deleuze's Spinozist account, the question of a life, characterised as a striving movement towards absolute potential and absolute beatitude, is a question of situation. Situation is where a mode of life, amid its everyday striving, discovers itself within its material circumstances. A mode of life is necessarily situated, and from its situation undertakes a survey of an immediate neighbourhood of opportunity and constraint.

Below I will describe a mode of life as the process of formation of a subject, that is to say, as an open *process of subjectivation*. Deleuze and Félix Guattari describe the activity of undertaking a survey across a plane of immanence in What is Philosophy?, where they explain that the plane of immanence, together with the construction of concepts, is 'needed to make up the "slow beings" that we are' (1994: 36). A mode of life, a specific subject – let's say for the purposes of this chapter, a creative practitioner like Margit Brünner whose work I will shortly introduce – undertakes a survey of a plane of immanence as an everyday activity of striving. This striving from the thick midst of a plane of immanence draws attention to the struggle to maintain an ethical way of life as an affirmative and ethico-aesthetic pursuit. As Deleuze argues, 'In Spinoza's thought, life is not an idea, a matter of theory. It is always a way of being, one and the same eternal mode in all its attributes' (Deleuze 1988c: 13), where 'way of being' also draws attention to one's ethical comportment. To this 'way of being' I would add the situation, the place, the landscape of encounters or the environment-world in which a mode of life strives. Every mode of life is explored *in situ*, and the immanent plane of survey can also be taken quite literally as pertaining to a landscape of encounters composed of ecological relations, what I call an 'environment-world'. Spinoza's Ethics, for Deleuze, outlines 'a typology of immanent modes of existence' (Deleuze 1988c: 23), and he goes on to suggest that 'there is only one term, Life, that encompasses thought, but conversely this term is encompassed by thought' (1988c: 14). To this can be added Deleuze and Guattari's insistence that the plane of immanence describes 'the image thought gives itself of what it means to think, to make use of thought, to find one's bearings in thought' (Deleuze and Guattari 1994: 37), and as such to affirm one's embodied situation.

This chapter is dedicated to joy and beatitude in relation to feminist spatial practices, with a focus on the work of a creative practitioner, the artist-architect

Margit Brünner. By reference to Brünner's work, and through a reading of Deleuze's reading of Spinoza's *Ethics*, I will discuss how the conceptual-sensory structure of 'beatitude' promises a refuge of sorts from experimental-experiential striving (*conatus*). This is a refuge from both sad and joyful passions, one that is fleeting and lacking in durability, as modes of life are also apt to plunge down again through Spinoza's three ethics, or levels of knowledge, as identified by Deleuze. Of course, to suggest that a mode of life 'plunges downwards' or rises upwards through the three levels of knowledge is to immediately suggest a hierarchy, which is not what I intend. Instead something elastic pertains to modes of life, inflated with joy, or deflated with sad passions, joy and passion manifesting as shifting proportions in relation to tentative, and sometimes more enduring, compositions, as life and thought are folded, unfolded, enfolded. Proportion here is less like a scale-bar of joys relative to sadnesses than like temperature plunging, or else atmospheric pressure rising, which might offer a better way of thinking through these transversally articulated alterations.

From a discussion of the process of diagramming experiential-experimental striving towards beatitude, I will proceed into a discussion of Margit Brünner's creative practice, and then I will further discuss Deleuze's reading of Spinoza's *Ethics*, which he describes as a journey across the three ethics, whereby a mode of life passes through the atmospheric media of shadow, colour, and light, from the first to the third level of knowledge, and finally towards a momentary state of beatitude.

Adventures in Diagramming towards Becoming Joy

To offer an account of the process of experimental-experiential striving, an emphasis can be placed on the diagrammatics of speeds and slownesses, lines and planes and bodies (E IIIPref.), which are mapped out according to geographical latitudes and longitudes that pertain to specific environment-worlds. Diagram is a Deleuzian and not a Spinozist term, and here it pertains to the journey taken by a mode of life, a process of subjectivation, as it aspires towards beatitude. What I propose is that in the fleeting glimmer of beatitude, where a mode of life achieves the greatest proportion of adequate ideas, a refuge or safe haven is provisionally created. This refuge provides the vantage point from which to observe-feel the infinite curve of the plane of immanence as it unfurls, to experience what Deleuze calls 'pure immanence . . . a life' (2001); a space of refuge within which a mode of life can rest briefly and gain some insight, as well as an outlook from their embedded, embodied situation. That is to say, while imagining an infinite survey, much like the one described by Deleuze and Guattari at the opening of *What is Philosophy?* across a 'plane of immanence' (1994: 20), I also want to insist on the locatedness and specificity of what it means for a mode of life to strive (*conatus*) in the 'vicinity of beatitude'.

In his reading of Spinoza's *Ethics* Deleuze privileges the fifth book, and it is here that this pause, this circumscribed refuge denominated as a state of beatitude, can be located. But a difficult terrain must first be explored, concerning what a 'given body seeks out and what it avoids', how it retreats from what harms it, and draws close to what aids it, which also concerns a 'particular being's characteristic relations with its surroundings' (Gatens and Lloyd 1999: 100). Where a 'particular being' begins to experience and experiment with their surroundings, a practical diagram emerges, forging a cartography that pertains to a specific journey. Moira Gatens and Genevieve Lloyd make an account of such dynamics in terms of ethological bodies where 'ethology describes the various powers of beings in relational terms by treating an individual as a fully integrated part of the context in which it lives and moves' (Gatens and Lloyd 1999: 110). It is this question of context and how a mode of life intimately interacts with a context as environmental milieu that I will foreground as I proceed, and which is exemplified, as I will explain further on, in Brünner's work.

In their joyful *and* sad wake, encounters between modes can be mapped out as lines that cluster geometrically to generate planes, producing bodies, or corporeal-ideational compositions, that can be surveyed according to an organisational matrix of latitude and longitude. Together, lines and planes and bodies, in motion and at rest, suggest a restless, dynamic, even seething 'transversal' movement as human and non-human bodies collide, recalibrate, combine, and decompose in relation to each other. A body, any body, is always to be found somewhere, at some time, and this question of situation should not be allowed to fall too far into the background. On the one hand, there are the qualitative measures of slownesses and speeds, or relations of motion and rest of bodies, which Deleuze describes under the geographical denominator of latitude. On the other hand, longitude accounts for the affects a body suffers or enjoys at any moment in light of the above physics of motion and rest. Deleuze respectively calls these modifications *vectorial signs*, which register the augmentation and diminution of the power of a body in relation to the circulation of affect (a vector suggests a directional movement, a journey); and *scalar signs*, which denominate the state of a body's affections at a specific moment, what a body is feeling right here and now, and whether its power is concurrently diminished or increased (Deleuze 1998: 140). Latitude and longitude, the physics of bodies together with the affects that are stirred in the dynamic interchange of affecting and affected, manifest as an orientation in thought-action that is place-based.

An extended quote from Deleuze's book *Spinoza: Practical Philosophy* helps to sketch out the organisational matrix of this diagrammatics of affected and affecting, and this curious cartography of latitude and longitude:

> In short, if we are Spinozists we will not define a thing by its form, nor by its organs and its functions, nor as a substance or a subject. Borrowing terms from the Middle Ages, or from geography, we will define it by longitude and latitude. A body can be anything: it can be an animal, a body of sounds, a mind or an idea: it can be a linguistic corpus, a social body, a collectivity. We call longitude of a body the set of relations of speed and slowness, of motion and rest, between particles that compose it from this point of view, that is, between unformed elements. We call latitude the set of affects that occupy a body at

each moment, that is, the intensive states of an anonymous force (force for existing, capacity for being affected). In this way we construct the map of a body. The longitudes and latitudes together constitute Nature, the plane of immanence and consistency, which is always variable and is constantly being altered, composed and recomposed, by individuals and collectives. (1988c: 127)

The above excerpt describes a cartography of affect that must be open to constant adaptation and renovation, following the dynamic transformations of a particular being or mode of life, which is never assumed to be a stable, self-same subject. Spinozists such as Warren Montag and Louis Althusser before him have complained of the great difficulties and challenges of reading Spinoza (Montag 1999: xiii), of making their way through the 'geometric method' of his *Ethics*. At the same time writers such as Deleuze seem less concerned with this difficulty, drawing out a cartography, as illustrated above, or else describing a wondrous river, branching into a thousand tributaries, slowing down, and speeding up (Deleuze 1998: 138). Deleuze describes a long, continuous, even grandiose and serene movement through Spinoza's *Ethics* via definitions, axioms, postulates, propositions, demonstrations, corollaries, and scholia. Yet he also warns that this first perception of a grand river is inadequate, even misleading. And after all, where must we commence but from amid inadequate ideas and signs referring to other signs obscured by shadows and half-light? Deleuze performs an entry into Spinoza's work, demonstrating how the reader must grapple with inadequacy in order to progress through what he calls the three ethics; Deleuze's idiosyncratic cartography of Spinoza's three kinds of knowledge. I place an emphasis on the cartography that Deleuze lays out, whether legitimately or not, across Spinoza's *Ethics*, because a spatial articulation of the *Ethics* becomes one useful means by which to inhabit its movements, one way of acknowledging an inextricable entanglement in a local environment-world.

The entire *Ethics*, Deleuze explains, 'is a voyage in immanence' (1988c: 29), and by the conclusion of his 'Spinoza and the Three "Ethics"', we have discovered the landscape features and encounters of his perhaps idiosyncratic reading. The first ethics (or kind of knowledge) he calls a river-book; the second (composed of Spinoza's notes or scholia) is subterranean, a coloured book of fire, and of laughter and hatred; the third ethics, finally, is an aerial book of light (1988c: 32). The three ethics progress across a series of thresholds according to sign, concept, and essence, which correspond to three specific media of shadow, colour, and light. Where the first kind of knowledge fumbles in the shadows, and scalar or else vectorial signs referring to other signs (*ad infinitum*) are aroused in the chance encounters between bodies, knowledge is gained only through the shadows that are so cast; the second kind of knowledge takes on modes as projections of coloured light procuring complementary relations (common notions) towards their full expression in white light (an infinite mode) (Deleuze 1998: 143). This full achievement would then be the third kind of knowledge, and its third state of light: 'no longer signs of shadows, nor of light as colour, but light in itself and for itself' (1998: 148). Importantly, this progression is not consecutive, leaving each step behind, but instead 'each of the ethics coexists with the others and is taken up in the others, despite their differences in kind. It is one and the same world. Each send out bridges in order to cross the emptiness that separates them' (1998: 33). This landscape composition and its cartographic features and media of shadow, colour, and light resembles the laying out of the plane of immanence, the 'planomenon' of Deleuze and Guattari's *What is Philosophy?*, where bridges are cast hopefully across the plane between concepts in order to articulate conceptual connections and answer the emergent problems of an environment-world (1994: 35–6).

Introducing the Architect-Artist Margit Brünner

Plunging down and rising up through the three ethics, or the three levels of knowledge, I will now draw on Margit Brünner's ethico-aesthetic labours. Through an engagement with Brünner's work, I suggest that beatitude manifests fleetingly in the rhythm of a sensory-conceptual-contraction-relaxation that creates a limited place or shelter from which further departures and returns can be provisionally planned. Her work offers more than a 'mere example' to illustrate a point: she is a mode *with whom* I, as something like a neighbouring mode, think-act. When I first encountered Margit Brünner she was falling out of a hammock while attempting to sketch a cluster of vibratory lines through the communicating pistil of a long prosthetic drawing device.[1] Brünner's work includes an impressive series of such quasi-functional instruments or tools to detect what she describes as local atmospheres of affect. She lost balance briefly, and tumbled to the floor with laughter. If you are as fortunate as Brünner then your ethical experimentation will achieve encounters that produce joyful affects, including laughter, and laughing together. For Brünner, beatitude is ventured through such performative acts, a process she calls 'becoming-joy'. She plans and also documents these performances by means of diagrams, so as to explore ways of 'manouvering atmosphere' (Brünner 2015: 8). Amid her artistic practice and her everyday life she places an emphasis on striving and action, conceived in a projective sense as a means of exploring existential territories of becoming that are specific to her immediate, local environment-world.

By following Brünner's work across non-human landscapes of becoming, I want to ask how a practical approach to constructing the greatest possible compositions, those that enable a greater proportion of active joys over reactive passions, can be tested. Brünner's feminist creative practice is exceptional in that she has actively undertaken her own reading of Spinoza in order to supplement her practical work; she has ventured from outside philosophy to undertake the difficult challenge of engaging with Spinoza (Brünner 2015), much as Bernard Malamud's protagonist does in *The Fixer* (Malamud 1966). I identify Brünner's as a feminist practice on account of its emphasis on critical embodied performance; how she explores alternative understandings of technology by way of her series of invented atmospheric tools; the attention she pays to relations between human and non-human things and places; and how her practice aims to procure 'situated knowledges' (Haraway 1988). Brünner's context concerns both environmental and disciplinary milieux, and in working along a transdisciplinary threshold and in close engagement with her surrounding environment-worlds she risks the exhaustion of her mode of existence as she fervently maps her ethological relations.

Brünner's work is ostensibly located between the spatial arts and performance art, but she is an architect. Her explorations endeavour to discover the best means of producing joyful affects, with an emphasis on the milieu, or the relationship between her local environment-world and her ever-transforming mode of life. This fundamental relation of mode of life as process of subjectification is an open project dependent on an ethological entanglement with environmental milieux. Importantly, the subject, whether she be a creative practitioner, a philosopher, or a child, is never a stable subject position given in advance, but a *process of subjectivation*, an emerging into being without ever resting in place, a *becoming*. In this regard, Deleuze argues that 'philosophy is a theory of multiplicities that refers to no subject as preliminary unity'; instead he replaces the idea of a pre-given or fixed and established subject with the notion of singularity (1991: 95). Supporting Deleuze's account of the process of subjectivation, Peter Pál Pelbart presents the formulation in this way: 'subjectivity as an event before a subject, an intensity before a form' (2015: 16). As Gatens and Lloyd also explain, ethological bodies are not concerned with pre-given moral ascriptions of 'type': they do not claim to know ahead of observation and experimentation. A mode of life, a process of subjectivation, undertakes experiments in relation to a milieu, and in the process of practical engagement a mode of life learns. Brünner's is such a practice of immanence, ever located, situated, and inspired by embodied learning.

In 2002, during her first visit to Australia, Brünner undertook a series of 'cosmethic space refinements', exploring methods for surveying and describing the atmospheres of specific public places, including the Docklands and Swanston Street in Melbourne. These urban landscape encounters followed investigations in urban areas and squares in the city of Vienna, such as Maria am Gestaade and the Donau-kanal. Brünner is Austrian, and while still in Austria she ventured into the urban outskirts of the Platte at the river Danube, and into the Viennese woods. In Australia she similarly complemented her urban landscape investigations with explorations of the landscapes of Brachina Gorge in the Flinders Ranges, Lake Mungo in northern New South Wales, and the wilderness around Broome in North Western Australia (Brünner 2015: 11 n.13). The invented tools she has tested for her surveys include catcher, surveyor, and pollination tools. The catcher is a tool that looks like a large, round, flat bat at the end of a long stick, which the practitioner waves about slowly, feeling the resistance between a human body and a body of air. The surveyor is the human body itself, clothed in padded white trousers and hooded top, standing stock still, frozen in place and hyper-attuned to its environment as crowds move about it. The pollination tool appears like a firecracker or a tube of fireflies that has been let off on a city street. Suffice to say, each tool is useless; it does nothing that appears functional, and yet each tool is perfect in its functionality as it achieves the role of mediator between one body and another, facilitating encounters with local environment-worlds.

Brünner's capacity to wield and manage her invented atmospheric tools is the means by which she explores the fundamental relation between a mode of life and an environmental milieu to see how far these encounters result in the modification of a composition of bodies. Between human and non-human bodies, between her own processes of subjectification and the urban and non-urban landscapes she encounters, and through the deployment of her tools, her aim is to produce greater compositions, acknowledging at the same time the ever-present risk of decomposition. Brünner explains the process of her 'cosmethics' (2011: 18 n.57), her means of extending a body through quasi-technical means, as the deployment of a 'surveying instrument' that extends reality. By this she means that she extends her grasp of what Spinoza calls 'adequate ideas' through attempts at understanding what she shares in common with her local environment-worlds. Her assemblages of mode of life *and* tool (thing) *and* environment-world 'communicate with the atmosphere, ever sifting, catching, memorising, absorbing, assimilating, transcribing, and translating. It is an active delayer, enlarger, intensifier, distiller, separator, catcher, stimulant and transporter of the emerging, fleeting, and growing phenomena', her aim being to 'reveal immanent moods and tempers'.[2]

The atmospheric composition so formed between self *and* thing *and* nature and/or God corresponds to what Deleuze describes as the three peaks of Spinoza's *Ethics*: Myself, Things, God. To orientate these peaks across an existential landscape, that is, amid a specific environment-world with which she hopes to engage, Brünner uses the central concept of 'atmosphere'. This term has been taken up somewhat uncritically by second-generation architectural phenomenology, in the writings of architects such as Peter Zumthor.[3] Brünner's atmospheres are operationally different because they are about creative practice. Her processes of atmospheric exploration necessarily include not just site-survey, but 'self-survey' (Brünner 2011: 14), and are always deeply embedded in specific contexts or milieux where processes of subjectification are apt to emerge. Atmosphere expands and contracts for Brünner as she tests its openings and delimitations with her conceptual and material toolbox, attempting to grasp what a body can do. Her engagements with non-human landscapes do not make a distinction between the natural and the cultural, but stress instead an approach driven by the urgent question: how do I enter into dialogue with my environment-world understood as affective atmosphere? She admits that the joy she pursues resists being utilised for representative purposes (Brünner 2011: 153). This can result in a failure of representational means, a limitation of our capacity to capture, through video, drawing, or photography, the profound encounter that has just taken place. She must make the best of the inadequate signs she uses for the purposes of documentation, so that a common notion might be shared with a critic such as myself, or with other readers of her work, and even a flash of intuition achieved.

Following the Material

Brünner's is an ethical experimentation in 'following the material' of local environment-worlds, and this is why she places an emphasis on her series of 'test-sites' (Brünner 2015: 8). Deleuze explains that existence itself can be conceived as a kind of test: 'Not, it is true, a moral one, but a physical or chemical test, like that whereby work[wo]men check the quality of some material, of a metal or a vase' (Deleuze 1992: 317; 1988c: 40). Brünner takes this notion of testing and applies it to specific sites. The Spinozist formula for Deleuze, as for Brünner, is 'We experience . . . we experiment' (Deleuze 2003: 1). This is not a question of forming judgements, though it can be a process of learning through encounters and the relations that endure following encounters. With respect to ethical experimentation, which is what the philosopher of science Isabelle Stengers calls an ecology of practices, it is crucial to point out a distinction between morality, which overdetermines our relations in a world through pre-given codes, and ethics as a practice worked out between transforming embodied processes of subjectification and a local, situated environment-world or milieu. Ethical experimentation (and the French word *expérience*) draws the terms experience and experiment together, and ethical experimentation suggests a way of following the materials of a situation. Brünner follows the materials of her encounters and hones her 'atmospheric skills' through a practical process of learning.

As Brünner explains in the 'glossary' of *Constructing Atmospheres: Test-sites for an Aesthetics of Joy*, atmospheric practice is a 'method of becoming joy' (2015: 49, 219). She follows the Spinozist formula of the proportional passage of affect: where sad passions reduce the proportion of a mode's capacities of expression, joyful affects empower a capacity to act in a world, increasing the proportion of active affects towards an affirmative difference. 'Ethology, whenever human practices are involved', as Stengers explains, following her reading of Deleuze on Spinoza, 'is based on productive, on performative experimentation with regard to modes of existence, ways of affecting and being affected, requiring and being obligated. . .' (Stengers 2010: 58). In fact Brünner dispenses entirely with the distinction between art and everyday practices and suggests that practice is about daily navigation towards making the best of all possible encounters: it's a tireless field-testing. Her cosmology is brought together with her ethics towards a joyful *cosmethics*, where ethology is less argued for than performed.

Brünner, from the midst of her practice, by following the materials immediate to her, asks: how can a mode of life achieve the greatest proportion of joyful affects towards the production of adequate ideas? Joy must be immediately distinguished from a more anodyne happiness, and it is certainly not bliss, which belongs to the passions. Joy is achieved through violent encounters and shocks to thought, at least according to Deleuze's interpretation of Spinoza. Brünner offers an alternative to such violence in the encounter with thought. Instead, she puts forward a feminist becoming with world, a soft swoon that speaks less of violence than of resonance and a shimmering vibration. It is most likely that we never or rarely achieve the greatest proportion of active joys and the least proportion of passive sadnesses, and when we do, like a journey, it must be considered a mountain peak of sorts. A moment, a vision, a shock, an opening, or – why not? – a swoon, an epiphany, and then we fall back again into stupor, stupidity, into the blur of mere inadequate signs, one after the next, as the productions of the imagination strive onwards.

A Journey through the *Ethics*

To describe this striving across non-human landscapes in more detail, I will return again to Deleuze's account, which proceeds as a journey through Spinoza's *Ethics* across the three ethics, the three kinds of knowledge. Our starting point is in a state of stupor – where else can it be? – confounded by signs that are inadequate, often obscured, and yet enlivened by the imagination. This is the first level of knowledge. This is where extrinsic encounters between bodies arouse inadequate ideas, images, and signs (Deleuze 1992: 303). These inadequate signs can be read as symptoms, as of a patient, or a passive recipient, one who is affected but produces little in the way of her own affects (Deleuze 1988c: 73, 74). At this level of knowledge all that can be ascertained is a general restlessness, a capacity for extension, for motion and rest, that is to say, a relation that can be demarcated according to latitude (1988c: 55). And here is where the imagination undertakes its labour, aroused in response to an imprint of some body made upon our own, a line that is drawn between me and you to form a fleeting or more resilient relation. Here is where memory begins to contribute to learning, where a succession of such imprints (Deleuze 1992: 311), a line drawn between now and then, reinforces a connection between past encounters between things. Just because imagination and memory are here with the passions at the first level of knowledge, this does not mean we should diminish their fundamental contribution, as Gatens and Lloyd strenuously argue in their book *Collective Imaginings* (1999). Spinoza acknowledges that the imagination can assist in the composition of wondrous things, even bodies politic, with the proviso that we understand that it is the imagination at work producing useful fictions that enable people to live and work together (Gatens and Lloyd 1999: 34–5). We could never commence or proceed without these passions and inadequate ideas, and we should never assume that we won't also return to them, as the fictions of the imagination can have practical uses and assist in the development of both political and creative practices.

Now from this state of passivity, if we are sufficiently receptive, and not entirely closed off to encounters, if we are open and prepared to be taken on this 'witch's ride', on this 'gust of wind' that is Spinozist (with all the risks that entails), maybe then, probably collectively, we might find or compose a common notion. A common notion is that which agrees with both my body and yours: you may be an ocean, and I may be a boat, or a swimmer. In any case, as Deleuze explains, 'properties common to our body and external bodies' (1992: 306) suffice to compose affective relations that increase our power of being, our capacity to persist as a mode of life. The goal of the second kind of knowledge is to 'overlay' the passive affects of the first kind of knowledge with active affects, and to keep in mind that passive affects and the stupor of inadequate signs, our fumbling trials and errors, and even our stupidities, might just hold the kernel of what could become an adequate idea (Deleuze 1988c: 60).

And as it turns out, there are peaks or summits following this progression through the three kinds of knowledge. Based on our agreements, and specifically where our compositions increase the capacity for expressions of life that can be carried resiliently forward, this is where adequate ideas promise to germinate. It is also crucial to point out that with the second kind of knowledge, joys of the second kind emerge. That is to say, we do not have to get to the third kind of knowledge to achieve joys, although joys of the second and the third kind are qualitatively and intellectually different (Deleuze 1992: 305–6). Where the former concerns shared concepts, the latter achieves flashes of intuition. With respect to joys of the second kind, a mode of life has not quite overcome all the sad passions, and has not quite achieved the greatest possible proportion of active affects. Where common notions are composed of colour, adequate ideas of the third kind are composed of pure light, as Deleuze elucidates (1998).

Let's say you make it to the third level of knowledge, qualified as intuition and expressed as pure light, then you would also be obliged to ask yourself: what kind of body am I now? The third kind of knowledge, with the greatest proportion of active joys and adequate ideas, and the near dissipation of the passions, does not yet mean that beatitude has been attained. You are not there yet (and you may well never be), but you have achieved joys of the second and third kind. And why would you want to go further? The univocity of substance in relation to a multiplicity of modes of a life, understood as expressed attributes, partakes in 'eternal essences' (Deleuze 1992: 313). Eternal essences are to be carefully distinguished from misleading notions of immortality. Deleuze explains that the object of the third kind of knowledge is to become conscious of God or Nature, myself, and other things (1988c: 61, 74). These are located as three summits, like mountain peaks, of the three levels of knowledge of the *Ethics*.

On arriving at the third level of knowledge, what a mode of life achieves is a state of 'beatitude', but not immediately. This is a challenging concept for a secular audience, except that the relation with God that is secured is not based on any anthropomorphic figuration, but is instead identified with nature: *Deus sive Natura*. This is the radical shock to established thought, the shake-up of a dogmatic image of thought that Spinoza attains when he writes his *Ethics*: he dangerously puts forward not a 'moral, transcendent, creator god' (Deleuze 1988c: 17) but a being with absolute immanence. Benedict or Baruch Spinoza also happens to have a given name that means blessed. He is a lens grinder, that is, he practises a material art, and he dies from inhaling glass dust. It is important that Deleuze spends the time he does on this biography, for through Deleuze we encounter a mode of life that is Spinozist. It is crucial to form a relation with such specificities, or else nothing is at stake and no risk is taken if we maintain a safe distance.

Let's say we arrive at the third kind of knowledge, expressed via an intuition that alerts us to the perplexing realisation that we have simply arrived at the place where we already and eternally are, in a relation with God or nature. Where 'We think as God thinks, we experience the very feelings of God' or 'We think as nature thinks, we experience the very feelings of nature' (Deleuze 1992: 308), where we think and act immanently in relation to our situation. We proceed to intuitive knowledge of our essence only from experiments, via ethical tests that are both practical and material, that is to say, in contact with an environment-world. So it would seem that we appear to reach the third kind of knowledge, only to find ourselves 'where we already were', with a relaxation of all striving and a quiet, slow dissolution into serene immobility.

To recap this ethical itinerary, where the journey must be undertaken in what Deleuze calls a 'strict order', we progress 1) from inadequate ideas that come to us as passive modes of life, and the passive affections that flow from such encounters; 2) towards the formation of common notions that suggest relations of sociability, but only by means of an active selection from the series of inadequate ideas that have been encountered, and only based on the careful choices we make. This effort of selection is 'extremely hard, extremely difficult' (Deleuze 1992: 145). The art of selection also pertains to the examples we use for a discourse such as the one unfolding in this essay, because what a thinker suggests is exemplary must be chosen with the greatest of care (Deleuze 1992: 301). This means, for instance, that we might suffer a passive affection, but if we select it for a composition and it works, through this effort, and this creative act of selection, we make it adequate and turn it into a concept, and the test becomes how well we can we share this idea as a common notion. 3) When we get to adequate ideas, then we can venture into the field of active joys and also loves, entering into the third kind of knowledge. It is a vain hope if we think we will achieve only these adequate ideas because 'we will always have passions, and sadness, together with our passive joys' (Deleuze 1992: 310). We can continue to strive, and we do this according to a milieu or environment-world that we must acknowledge is qualified through duration and extension, mapped according to latitudes and longitudes, slownesses and speeds or the relational capacities of modes, and what affects are exchanged as we go. The implication of this is that although Deleuze insists on a strict order through the levels of knowledge, any mode of life necessarily moves backwards and forwards, contracting and expanding with respect to their specific capacity for a life. Just because I contract to myself an adequate idea and a joy today, this is no guarantee that tomorrow I will not be all sad passions, and as such a rhythm of modes of life and their expressive capacities comes to be expressed. The imperative remains: I must continue, experiencing-experimenting, and following the materials of my worldly encounters.

The struggle continues, and even as we strive as much as possible to reduce our passions where they risk decomposing us, and remember that 'we' may be a body of water, a body of thought, a body politic, we cannot eliminate all passion during our existence. As Deleuze explains, our extensive parts are determined from outside *ad infinitum*. This is the extensive fact of our encounters, and mixtures of bodies (Deleuze 1992: 311; 1998: 142): as bodies we are necessarily open to a relative outside or environment-world; this is the mode of life in its enduring existence. To move beyond merely random encounters, Deleuze extends an imperative: 'we must in our existence come into what is our essence' (1992: 307). Essence is the determined nature of a thing, that which determines its capacities and its power to persist. As distinct from existence, essence has no duration, it is 'eternal'. Deleuze goes on to explain that essence is a degree or power or intensity; it expresses itself in our relations, but is not itself a relation experienced through time (1992: 312), that is to say: 'essence does not endure'. It has no duration, it is non-durable, and it cannot stand up on its own, and yet it has reality or existence *sub specie aeternitatis*, that is to say, from the perspective of the eternal. Yet rather than being universally and eternally true, this point of view is immanent, located in the right here and right now. As Deleuze argues, this is not a situation of being immanent *to* something or someone, but a situation of pure immanence, existing in the immediacy of a moment. He argues that 'In Spinoza, immanence is not immanence *to* substance; rather substance and modes are in immanence' (Deleuze 2001: 26). The implication is that essence relies on existence, in this immanent moment, and yet existence can strive no more without essence, understood as the particular set of determined capacities of a specific mode of life in relation to its environment-world. The cessation of struggle is suddenly upon us, it is immanent, immediate, locating a moment where no further increase in power is possible or relevant, and the greatest proportion of adequate ideas means a state of blessedness is achieved. Blessedness clears out a reserve, a shelter of sorts. This is because beatitude, or blessedness, no longer implies transitions and passages across the three levels of knowledge. A state of beatitude is no longer explained by duration; instead, it is the formal possession of a power of perfection (Deleuze 1988c: 51). This is how the encounters, selections, and transitions all the way to beatitude work: 'while the body exists, duration and eternity co-exist' (Deleuze 1992: 314), which means, according to Deleuze, that 'the soul's eternity can be indeed the object of a direct experience' (1992: 314). According to this formula, which involves proportion as a measure, 'the good or strong individual is the one who exists so fully or so intensely that he has gained eternity in his lifetime, so that death, always extensive, always external, is of little significance to him' (Deleuze 1988c: 41).

When we pause, as Brünner does, to ask where we are on the spectrum between joys and sadnesses, between activity and passivity, this suggests the nature of the diagram we are participating in drawing out, as well as a proportional measure. A diagram sketches out a body and/or thought's capacities, to affect, to be affected, and how a mode of life emerges from scalar and vectorial signs, mapped out according to latitude and longitude, towards concepts and even flashes of intuition. This also pertains to how a mode of life, a process of subjectivation in relation to an environmental milieu, can achieve sufficient consistency and durability, at least for the time being; that is, before going on to suffer and enjoy further adventures in affect, whereby a mode of life becomes yet further transformed.

Proportion includes not just the expansions and contractions of how a mode of life affects and comes to be affected, but how entanglements between subjects in the midst of transformation and environment-worlds are mapped. Proportion is an expanding and contracting measure describing the dynamic momentum of the motion and rest of bodies (inclusive of human, landscape, and other bodies). Without a prospective end in sight, this dynamic momentum instead becomes a learning process that lends itself to future practice-based experiments, and this is what is of significance in the brief account I offer of Brünner's work. Her engagements with specific environment-worlds understood as specific situations do not aim at a final project that can be neatly circumscribed and classified. Instead her work draws attention to an ongoing and open project, ever at risk of exhaustion, as well as alerting us to the risk of the exhaustion of the environment-worlds in which we are embedded. We must always commence from somewhere, and most often that somewhere pertains to our local environment-world with which we are co-constituted. So why not start where we are, with those problems that would appear to be most immediate, and most pressing? To address problems that are immanent to our fields of action means developing an adequate and ethical practice of a life, where the 'ethical test is therefore the contrary of the deferred judgment: instead of restoring a moral order, it confirms, here and now, the immanent order of essences and their states' (Deleuze 1988c: 41), which is to say that proportion does not allow a mode of life to rest easy; instead it continues to increase and diminish according to immanent and transformative situations.

Notes

1. I first met Margit Brünner at a symposium and workshop entitled 'Expanded Writing Practices' at the University of South Australia in September 2009, which brought together practitioners and theorists from art and architecture, and was organised by Linda Marie Walker and John Barbour.
2. Margit Brünner, email correspondence with the author, 24 September 2013.
3. It is also a term that has gained currency in relation to the work of thinkers such as Gernot Böhme, Peter Sloterdijk, and Ben Anderson.

11

The Eyes of the Mind: Proportion in Spinoza, Swift, and Ibn Tufayl

Anthony Uhlmann

This chapter will examine the interconnected ideas of proportion and relation in Spinoza by reading Spinoza alongside two novels that have been drawn into proximity with his works.[1] The first is a 'Spinozistic novel' *avant la lettre*, written by the twelfth-century Islamic philosopher Ibn Tufayl, which was translated by a friend of Spinoza's because of its resonance with his philosophy. This novel, I will argue, sheds light on the importance of relation and the absence of relation to Spinoza's three kinds of knowledge. By contrast, the second novel, Jonathan Swift's *Gulliver's Travels*, seeks to criticise Spinozism, and at times Spinoza (see Gardiner 2000), in part by considering the idea of proportion and its connection to reason. To begin, however, I will sketch a background to some of the claims concerning Spinoza's effect on English literature in the seventeenth and early eighteenth centuries.

The Radical Enlightenment and English Literature

In *Radical Enlightenment* and *Enlightenment Contested*, Jonathan Israel offers a major rethinking of the importance of Spinoza's ideas and the effect of his works on the European Enlightenment from 1650 to 1752. Israel recovers the central role of Spinoza in the processes that produced the Enlightenment, a role that he argues had been obscured because it had to take place underground. During the period of study, it was more or less forbidden to mention Spinoza's works, or ideas that were seen to align with them (called 'Deism' in England, or, following Israel, Spinozism), directly, unless one was repudiating them out of hand. The ideas that characterised Spinozism included the denial of a providential God, the identification of God with nature, and the insistence on seeking truth through reason rather than through reference to authorities such as the Church or the Bible. Even those who sought a more nuanced critical engagement with Spinozism had to be very careful. In effect, the only possibility was to engage with it indirectly. Israel states:

> By the early eighteenth century the widening perception of Spinozism as the prime and most absolute antithesis and adversary of received authority, tradition, privilege, and Christianity had generated a psychological tension

evident throughout the academic world and 'Republic of Letters', not unlike the intellectual and ideological paranoia regarding Marxism pervading western societies in the early and mid-twentieth century. To label someone a 'Spinozist' or given to Spinozist propensities was effectively to demonize that person and demand his being treated as an outcast, public enemy, and fugitive. Conversely, for an academic, court savant, official, man of letters, publisher, or ecclesiastic to be publicly decried as a 'Spinozist', or privately rumoured to be such, constituted the gravest possible challenge to one's status, prospects, and reputation, as well as standing in the eyes of posterity. (Israel 2001: 436)

The analogy between Spinozism and Marxism is illuminating. Just as Marxism could not have existed without the works of Karl Marx, those repudiating it, and even at times those affirming it, did not necessarily engage directly with those works. Those affected by Marxism, just like those affected by Spinozism, were often not directly acquainted with the works of Marx or Spinoza; rather, crucial elements of their ideas had been assimilated by others.

While Israel does not focus on English literary history, or the concerns developed in it, the information he offers underpins an understanding of elements of Spinozism's impact on it. The first major controversy involving Spinozist thought was the anonymous publication in 1670 of his *Tractatus Theologico-Politicus*. This was immediately banned but published clandestinely, with various tricks used to confuse the authorities as to the author, place, and date of the book, since its publication was deemed illegal soon after the first edition appeared. An edition known as the 'English Edition' was published (in Latin) with an English-styled title page in 1674, in order to be exported to England. Israel indicates that the initial reception of Spinoza in England can be traced to 1674–76 (Israel 2001: 275–85). While the *Tractatus* was a banned book, it was 'everywhere known' among the intellectual elites, at least by reputation, because of the then shocking challenges it posed to received theological and political ideas. Israel shows how English thinkers such as Locke, Boyle, and Newton responded to, and attempted to come to terms with or develop alternatives to, Spinoza's work (Israel 2001: 252–7).

In the *Tractatus* Spinoza uses reason to analyse the authority of scripture and to challenge the religious authorities' requirement that reason accommodate itself to the revealed truth of the scriptures. Spinoza contends that theology and philosophy are separate and that each should keep to its own domain (TTP ch. 15/G III 180). He further contends that free thinking is necessary to philosophical reflection and should always be allowed, with men judged for their deeds, not their thoughts or words. True philosophy is not antithetical to God or impious (TTP ch. 14/G III 173–80). For Spinoza, the principal function of religion is to affirm two fundamental truths: first, that God exists and can be known, and second, that we should love our neighbours as we love ourselves. The teachings of the scriptures all lead to this (TTP ch. 13/G III 167–72).

The ideas of the *Tractatus* were surprisingly influential in Britain and Ireland. With regard to this reception, Israel states:

It has perhaps never been sufficiently emphasized that in England and Ireland, where intellectual debate unfolded within a predominantly national context sometimes tinged with xenophobia, and with very few foreign writers being regularly cited, a pervasive, even at times obsessive, preoccupation with Spinoza persisted from the mid-1670s throughout the rest of the early Enlightenment. Spinoza and his books were indeed discussed by an extraordinarily large number of British and Irish writers, including – leaving aside the deists – key scientists, such as Boyle and Nehemiah Grew, university dons such as Henry More, Ralph Cudworth and Richard Bentley, and churchmen of many hues. (Israel 2001: 599)

Rosalie Colie remains the principal source on how Spinoza impacted on English literature in general terms (Colie 1959; 1963), though a number of others have touched upon this with regard to individual authors (some examples are listed below). The clandestine nature of Spinozism may have meant that its effects have been rendered virtually invisible to contemporary literary historians. That is, except for a few radical Deists such as Toland, Tindall, and Clarke, it has been widely believed that English intellectuals only responded to Spinoza obliquely and indirectly, whether they were seeking to express agreement (as perhaps Rochester and Pope do)[2] or to answer the challenge of his ideas (as perhaps Defoe, Dryden, and Swift do).[3]

In short, a list of canonical writers seem to have been responding to Spinozism from the very beginning of his reception in England right through the development of the novel in English, prior to the strong 'rediscovery' of Spinoza that took place in German and English Romanticism (Bell 1984; Hooton 1999). Part of my contention, then, is that the conditions that led to the emergence of the novel in English (and the development of the novel more generally in Europe) need to be reconsidered in light of the effects of Spinozism and with reference to Spinoza's works.

Ian Watt, among others, underlines the importance of causal thinking, or lines of causation, to the development of the realist mode in fiction, which is usually felt to begin with Defoe's *Robinson Crusoe* in 1719 and is forcefully aligned with the emergence of the novel as a dominant form in English. Watt claims that

> Modern realism . . . begins from the position that truth can be discovered by the individual through his senses: it has its origins in Descartes and Locke, and received its first full formulation by Thomas Reid in the middle of the eighteenth century. (Watt 2001: 12)

Like most other critics, Watt ignores Spinoza here, even though Spinoza moves beyond Descartes in extending causal determination to the mind as much as to things, and even though Spinoza's understanding of causation is far more complex than that set out by Locke. Watt may pass over Spinoza because, like many others, he equates the new mode of realism in the novel with British empiricism, which some have seen to be in opposition to Spinoza.

The early supporters of Spinoza turned to the novel form itself to spread some of his key ideas, developing, in both French and Dutch, what Israel calls the Spinozistic novel, beginning with the Dutch novel *Philopater* (Israel 2001: 315–19, 591–8). Similarly, the twelfth-century Islamic philosopher Ibn Tufayl's proto-novel *Hayy Ibn Yaqzan*, which I will discuss below, was taken as representative of an Islamic wisdom and tolerance of reason.[4] A further example of how fiction might allow an indirect representation of Spinozistic ideas is *Lettres persanes* by Montesquieu, who, like Pope, was forced to refute the charge that he was a Spinozist. Israel underlines how Islam or Persia, when used in works of fiction, became a kind of code for Spinozism (Israel 2006: 615–39). Confucianism was also read as affirming some principles of Spinoza's (Israel 2006: 640–62). All of this affirms that an indirect approach was seen to be beneficial for those who wanted to speak in praise of Spinozist ideas. Even those, like Defoe and Swift, who wished to engage with or refute Spinozism, worked in an indirect manner.

Relation and Proportion in Spinoza and Ibn Tufayl

On 26 November 1669 eleven men met in Amsterdam to form an 'academy' called *Nil Volentibus Arduum* ('Nothing is difficult for those with sufficient will'), which sought to bring together philosophy, the sciences, and the arts, and to reformulate Dutch theatre on the model of French theatre. Two of them, Johannes Bouwmeester and Lodewijk Meyer, knew Spinoza well and corresponded with him (van Suchtelen 1987: 391). Van Suchtelen sets out how the group as a whole were 'friends of Spinoza' and considered him their intellectual guide, and how the activities they undertook promoted the ideas developed in his philosophy (1987: 392–404). They met every Tuesday from 5pm until 8pm, until Meyer's death in 1681 (van Suchtelen 1987: 393). One of the tasks that *Nil Volentibus Arduum* set themselves was to educate the Dutch people and help them to better follow the light of reason as recommended in Spinoza's philosophy and philosophical rationalism more generally. They set about doing this by writing dictionaries and pamphlets, but one of their main activities was the translation of works of literature (theatre in particular) that they considered edifying, opening a general public to the light of reason.

At their meeting of 29 December 1671 the group requested that Bouwmeester translate *Hayy Ibn Yaqzan*, a novel written during the Islamic enlightenment in twelfth-century Spain by the Muslim philosopher Ibn Tufayl (van Suchtelen 1987: 397). This work had been translated into Latin by Edward Pocock in 1671 and Bouwmeester worked from the Latin in developing his Dutch translation (Israel 2006: 628).

The novel itself states that its purpose is to make the universal light of reason available to all, and indicates that such knowledge is only obscured by our prejudices. Bouwmeester esteemed Ibn Tufayl for showing 'how someone can, without any contact with other people, and without education, arrive at knowledge of himself, and of God' (Israel 2006: 629). Bouwmeester also seemingly had some role in the production of Spinoza's *Opera posthuma*, and Spinoza had written to him suggesting that he might translate the *Ethics* from 'part 3 onwards', though he did not, in the end, do this (Steenbakkers 1994: 16–17). While it is not possible to establish with certainty that Spinoza read Ibn Tufayl, given his close association with Bouwmeester and Meyer it is quite plausible that he did. Even if he did not, the resonances between *Hayy Ibn Yaqzan* and Spinoza's system serve to explain the enthusiasm of *Nil Volentibus Arduum* for the novel. Further, as I will set out below, a comparison of Spinoza's *Ethics* and this novel brings to light the importance of relation (and proportion) in Spinoza's work.

The works of Ibn Tufayl and Spinoza can best be drawn together around the idea of relation and the absence of relation and the central roles these play in their systems. Spinoza's philosophy is built around the Latin *ratio*, a word that means at once reason, proportional relation, and relation. In turn each of the three kinds of knowledge is shown to involve relation as part of its essence. This is underlined when Spinoza distinguishes the three kinds of knowledge by referring to an example to explain each kind. This example concerns our possible understandings of the principle of proportional relation set out in Euclid.

> Suppose there are three numbers, and the problem is to find a fourth which is to the third as the second is to the first. Merchants do not hesitate to multiply the second by the third, and divide the product by the first, because they have not yet forgotten what they heard from their teacher without any demonstration,[5] or because they have often found this in the simplest numbers, or from the force of the Demonstration of P7 in Bk. VII of Euclid, viz. from the property of proportionals. But in the simplest numbers none of this is necessary. Given the numbers 1, 2, and 3, no one fails to see that the fourth proportional number is 6 – and we see this much more clearly because we infer the fourth number from the ratio which, in one glance, we see the first number to have to the second. (E IIP40S2)[6]

Here the first kind of knowledge proceeds from the association of ideas (inexact relations) through which the merchants associate the answer with the technique they learned from their teachers. The second kind of knowledge proceeds through the understanding of exact mathematical or logical relations that allow the generation of solutions to problems, the demonstration of which is set out in Euclid. The third kind of knowledge involves immediate recognition of a relation, which in E IIP40S2 is a proportional relation or ratio.

The central character of the novel *Hayy Ibn Yaqzan* [Life, son of Awake or Consciousness] comes to his island in one of two ways, about which the narrator claims to be unsure: either he came ashore in a basket as an infant or was formed direct from the earth.[7] In any case he is found by a doe, who pities him and adopts him as her own. Without human contact or even human language, the natural light of reason allows him to attain first rational, and then spiritual excellence, with the former helping him to attain the latter. As he matures and seeks greater perfection, he looks to imitate God, who he has come to understand through natural necessity. To do this he identifies three kinds of imitation which resonate with Spinoza's three kinds of knowledge.

The three kinds of knowledge for Spinoza are the imagination, scientific or logical reason, and intuition. Like Ibn Tufayl's kinds of imitation they ascend from the most imperfect to the most perfect. For Ibn Tufayl, there are three kinds of imitation that will lead to a communion with God. The first is the imitation of animal forms. While, like Spinoza's imagination, this can lead one astray or into error it remains important, as we have been given bodies and need to think through and with them. The second kind of imitation relates to a contemplation of the mind or consciousness, which is in proportion to the spiritual realm, which is further reflected in the stars and planets. That is, he talks of imitating those heavenly bodies because our minds are not of the body but of some spiritual being, which is further reflected in the heavenly bodies. Speaking of the mind he states:

> This conscious part was something sovereign, divine, unchanging and untouched by decay, indescribable in physical terms, invisible to both sense and imagination, unknowable through any instrument but itself, yet self-discovered, at once the knower, the known and the knowing, the subject and object of consciousness, and consciousness itself. (Ibn Tufayl 2003: 141–2)

If the first imitation contemplates the world, or things perceived, the second contemplates the consciousness that comprises the self, or things known. This might be compared to Spinoza's second kind of knowledge or ratiocination.

The third imitation for Ibn Tufayl involves the contemplation of God himself. Here his definition of God, as the one whose existence is necessary, is exactly aligned with Spinoza's definition of God or nature, which alone has an essence that involves existence. To quote Ibn Tufayl:

> The third sort of imitation is attainment of the pure beatific experience, submersion, concentration on Him alone Whose existence is necessary. (Ibn Tufayl 2003: 143)

Spinoza's third kind of knowledge, intuition, is similarly tied to a direct knowledge of God, or the existence of God, nature, or existence itself, as that which occurs to us most forcefully as true. For Ibn Tufayl, this is a purely spiritual endeavour and links to a practice akin to, or identical with, meditation. One contemplates the stages of imitation in order to pass away into the One.

Yet with the processes described by Ibn Tufayl there is a significant difference from Spinoza. Rather than achieving understanding directly through reason as in Spinoza, for Ibn Tufayl such understanding is achieved through contemplation or meditation. What is most striking is that the difference underlines how the spiritual dimension described by Ibn Tufayl involves an *absence* of relation, or comparison, or proportion. This is true of the second kind of imitation. To take up again and complete a previous quotation, he states:

> [Consciousness is] unknowable through any instrument but itself, yet self-discovered, at once the knower, the known and the knowing, the subject and object of consciousness, and consciousness itself. There is no distinction among the three, for distinction and disjunction apply to bodies. But here there is no body and physical predicates and *relations do not apply*. (Ibn Tufayl 2003: 141–2, my italics)

This radical lack of relation is insisted upon again with the third kind of knowledge. It is worth repeating and continuing the passage cited above regarding the third kind of imitation. He states:

> The third sort of imitation is attainment of the pure beatific experience, submersion, concentration on Him alone Whose existence is necessary. In this experience the self vanishes; it is extinguished, obliterated – and so are all other subjectivities. All that remains is the One, True identity, the Necessarily Existent – glory, exaltation, and honor to Him. (Ibn Tufayl 2003: 143)

Tellingly then, ratio (relation or proportional relation), which allows thought in the first and second kind of knowledge in Spinoza, 'does not apply' in the second imitation of Ibn Tufayl and is absent or replaced by 'True identity' in the third kind of imitation. While there is insufficient space to consider the problem here, this kind of absence of relation might allow a reading of Spinoza's third kind of knowledge which might also be understood through the absence of relation rather than relation itself.[8] That is, it does not involve any comparison or proportional relation.

Yet there is another angle from which these things can be viewed in Spinoza, one that takes us in a different direction from this incorporeal understanding of the indivisible One or substance. Spinoza states in Proposition 23 of Part 5 that 'The human mind cannot be absolutely destroyed with the body, but something of it remains which is eternal' (E VP23). This difficult idea is extended in the scholium to this proposition where he talks of 'the eyes of the mind':

> There is, as we have said, this idea, which expresses the essence of the body under a species of eternity, a certain mode of thinking, which pertains to the essence of the mind, and which is necessarily eternal. And though it is impossible that we should recollect that we existed before the body – since there cannot be any traces of this in the body, and eternity can neither be defined by time nor have any relation to time – still, we feel and know by experience that we are eternal. For the mind feels those things that it conceives in understanding no less than those it has in the memory. For the eyes of the mind, by which it sees and observes things, are the demonstrations themselves. (E V23S)

What is striking here is that, in seeking to explain the third kind of knowledge, and the intuition of our own eternal essence which can only be revealed through the third kind of knowledge, Spinoza turns to a figure that relates to the first kind of knowledge: perception, feeling. It is not possible simply to dismiss this as an analogy, because we only understand *that* we understand through such feeling, 'for the mind feels those things that it conceives in the understanding'. In some way, then, the first kind of knowledge is being related to the third kind. So too, the body is implicated in the eternal ('the essence of the body under a species of eternity'). While Ibn Tufayl's eternal seems purely spiritual, then, for Spinoza, the essence of our bodies persists in relation to the eternal substance. I will not pretend to be able to unpack this here, but will merely point to this paradoxical relation.

Gulliver's Travels and Proportion

If Ibn Tufayl's work engages with Spinoza's to shed light on the importance of relation and the absence of relation to the philosophical systems developed by each, Jonathan Swift's *Gulliver's Travels*, which criticises Spinozism and other elements of the new philosophy, engages with Spinoza with regard to the nature of proportion. Indeed, Swift's counter-understanding of proportion might serve to draw Spinoza's system into question.

Gardiner (2000) has furnished a good deal of evidence for Swift's engagement with Spinoza, and Montag (1994) has written at length about the value of thinking of Swift in relation to Spinoza. Swift had Spinoza's complete works in his library and closely annotated the *Tractatus* (Gardiner 2000: 233), and he even wrote favourably of Spinoza, at least in comparison with his treatment of contemporaries such as Toland, Coward, and Tindall, and other English Deists who were also called Spinozists. In 'Remarks on a Book Entitled "The Rights of the Christian Church"' (1708), directed against its author Matthew Tindall, Swift writes:

> And truly, when I compare the former enemies to Christianity, such as Socinus, Hobbes, and Spinosa, with such of their successors, as Toland, Asgil, Coward, Gildon, this author of the Rights, and some others, the church appears to me like the sick old lion in the fable, who, after having his person outraged by the bull, the elephant, the horse, and the bear, took nothing so much to heart as to find himself at last insulted by the spurn of an ass. (Swift 1861: 254)

Gardiner shows how the figure of the horse, at the centre of Part 4 of *Gulliver's Travels*, 'A Voyage to the Country of the Houyhnhnms', was connected in the period to philosophical atheism, and further directly to Spinoza. Gardiner shows how Gulliver's implicit maxim in Part 4, 'Be ye as a horse', contradicts Psalm 32.9, with its maxim 'Be ye not as a horse, or a mule, which have no understanding', suggesting that linking 'the atheist to the horse ... was common in the 17th century' (Gardiner 2000: 230) and that Spinoza was commonly represented at that time (however erroneously) as the atheist philosopher *par excellence*. In *Spinoza Reviv'd* (1709) William Carroll argued that, according to Spinoza's principles, a man and a horse were not on separate rungs of the chain of being

(Gardiner 2000: 232). Gardiner goes on to underline how the figure of the horse as a noble creature needing to be emancipated from its rider (where humanity by analogy was linked to the horse and the clergy to the rider) is central to Matthew Tindall's book *The Rights of the Christian Church*. As the above quotation makes clear, Swift knew Tindall to be strongly influenced by Spinoza. Indeed, researching the essay that challenged Tindall's book, Gardiner argues, seems to have led Swift to read Spinoza closely (2000: 232–3).

The idea of contrast as a method for analysing the received views and mores of the times in Europe is at the heart of *Gulliver's Travels*, and this contrast is often connected to the idea of proportion by Swift. Whereas proportion is forcefully linked to reason in Spinoza, in Swift the *absence* of proportion (as the general human condition) or an overemphasis on proportion (which Swift criticises in philosophy) are the object of satire. In Part 2 Gulliver goes to Brobdingnag, a land of giants, after visiting Lilliput, a land where he is the giant. In each case the proportional difference in size is 12 to 1:

> as human Creatures are observed to be more Savage and cruel in Proportion to their Bulk, what could I expect but to be a Morsel in the Mouth of the first among these enormous Barbarians that should happen to seize me? Undoubtedly Philosophers are in the Right, when they tell us that nothing is great or little otherwise than by Comparison. It might have pleased Fortune to let the *Lilliputians* find some Nation, where the People were as diminutive with respect to them, as they were to me. And who knows but that even this prodigious Race of Mortals might be equally overmatched in some distant Part of the World, whereof we have yet no Discovery? (Swift 1958: 62)

Here, for the moment, philosophers are in the right, in thinking in terms of comparison. Contrast or comparison is a method for opening the mind, and Swift, working in the satirical mode, uses it to offer critiques, at once savage and hilarious, of contemporary England and Europe.[9]

Philosophy too is given over to critique, again through the use of contrast and comparison. The requirement for readers is to weigh the differences and identify absurdity. The absurdity is often linked to the figure of proportion, in this case both its absence (when philosophers lose contact with reality) and its over-abundance (when philosophers concentrate too heavily on proportional relations, only seeing pure mathematical forms and thereby failing to understand the less pure forms of human existence).

Comparison itself considers the proportion or relation between two things. The very small Lilliputians have ridiculously grand and pompous ideas but are small-minded and willing to utterly destroy their neighbours to glorify themselves (in comparison to them Gulliver is enlightened and merciful). The very large Brobdingnagians are hideous through the magnification of their physicality but are large in spirit, or at least their king is, refusing with horror Gulliver's offer to reveal the secrets of gunpowder and modern warfare; to them in comparison Gulliver is a moral insect.

Modern philosophy in general is a major target for satire in *Gulliver's Travels*. Part 3, 'A Voyage to Laputa, Balnibarbi, Luggnagg, Glubbdubdrib, and Japan', is largely given over to a comical critique of empirical philosophy and philosophical speculation, with the jokes again turning around the absence of proportion. The ruling class on the floating island of Laputa focus on mathematics and music to the neglect of all other disciplines or pursuits, and suffer in comparison to the so-called stupid and the uneducated, who seem infinitely wiser in their practices and ideas. In this case the philosophers are unreasonably preoccupied with proportion and ratio, which dominate the disciplines of mathematics and music. On the other hand, the 'Learned Academy of Practical and Speculative Science, Philosophy and the Arts' is given over entirely to grotesquely impractical pursuits; philosophy here is criticised for lacking proportion.

While Spinoza is not the sole object of Swift's critique, that he is one of the intended objects of it becomes apparent through references to Holland and Japan. Of those philosophers Swift mentions in 'The Rights of the Christian Church', only Spinoza was Dutch. Gardiner too underlines this point, suggesting that 'Swift creates a Dutch context for the last voyage to underscore that the new atheism emanates from Holland, where Spinoza still had a lively following' (Gardiner 2000: 234). Swift refers to 'New Holland', which is close to Houyhnhnmland, as is 'De Wits Island' (referring to the Dutch ruler Jan de Witt, whom Spinoza admired).

Yet Gardiner does not mention another point of connection to Spinoza. The Dutch agreement with the Japanese not to practise Christianity is described with disgust in Part 3 of *Gulliver's Travels*, where Gulliver has two bad encounters with Dutch sailors in the company of the Japanese at the beginning and end of the Part. This agreement is mentioned twice by Spinoza in the *Tractatus Theologico-Politicus*, with Spinoza then linked with his Dutch countrymen in outraging Christian sensibility. Spinoza writes:

> Concerning Christian ceremonies, namely baptism, the Lord's supper, feast-days, public prayers, and any others that are and always have been common to the whole of Christianity – if they actually were instituted by Christ or the Apostles (which is still not clear to me), they were instituted only as external signs of a universal church and not as things that contribute to happiness or have any sanctity in them. Hence, although these ceremonies were not instituted for the purpose of [upholding] a state, they were instituted only for a community as a whole. Consequently, a man living alone is not bound by them, and anyone who lives under a government where the Christian religion is forbidden is obliged to do without them and yet will be able to live a good life notwithstanding. We have an example of this in the empire of Japan, where the Christian religion is forbidden and the Dutch who live there, must abstain from all external worship by command of the [Dutch] East India Company. (TTP ch. 5/G III 76)

The point here is not simply that Spinoza and Swift mention the same case; rather, it has been established that Swift read closely and annotated his copy of

Spinoza's *Tractatus* while building his defence of Christianity against Tindall's seemingly Spinozist assault upon it (Gardiner 2000: 233), and that Swift was well aware of Spinoza's understanding of the practice of the Dutch in Japan. It is telling that Swift's reading offers a strong criticism of Spinoza's. Whereas Spinoza sees the agreement to renounce Christianity in a place where the religion is forbidden to be clearly justified on pragmatic grounds, for Swift, this renunciation – whose emblem was the requirement placed upon the supposedly Christian Dutch traders in Japan to trample upon the crucifix – is a blasphemous outrage. Further, in Part 4 it is noted that the horses, when they speak, sound as if they are speaking in 'High Dutch or German', calling to mind either Dutch philosophers such as Spinoza and perhaps Descartes, who resided there, or German philosophers such as Leibniz and Wolff (Swift 1958: 190).

Part 4 of *Gulliver's Travels*, 'A Voyage to the Country of the Houyhnhnms', also turns on questions of comparison and proportion. Swift does not dispute that a life led according to the tenets of pure reason such as that recommended by Spinoza and others would be good and desirable; rather, he contends that it would simply be impossible, because such a regime cannot be adapted to human nature, which is incapable of it. This is adequately demonstrated by Gulliver who, rather than developing wisdom through his contact with the Houyhnhnms (who, unlike any race of humans, conduct themselves purely and unerringly through the light of reason alone), becomes insufferably pompous and superior:

> *Yahoo* as I am, it is well known through all *Houyhnhnmland*, that, by the Instructions and Example of my illustrious master, I was able in the Compass of two Years (although I confess with the utmost Difficulty) to remove that infernal Habit of Lying, Shuffling, Deceiving, and Equivocating, so deeply rooted in the very Souls of all my Species; especially the *Europeans*. (Swift 1958: xxiv)

That human nature is ill-adapted to living in accordance with reason is further demonstrated by the exaggeration of the animal nature of human beings through the Yahoos, who, while they look exactly like humans, if perhaps more covered in filth than most, live in trees, grunt rather than talk, and throw excrement at those who offend them (among other odious habits).[10] Yet there are many sections of Part 4 in which the Houyhnhnms are used to contrast with the mores of Europe. They offer examples of good governance and working through the light of reason that strongly contrast with the kinds of governance witnessed in Europe and England, presenting the latter in an ill light. Unlike Part 3, then, where certain tendencies in experimental and speculative philosophy are held up for ridicule, in Part 4 the moral philosophy of the Houyhnhnms is not ridiculed because of the ideals on which it is based, but because it is seen to be impossible for human beings to put into effect. This is ironically demonstrated by Gulliver's disgust at the impossibility of reforming humanity even when the light of reason clearly points to areas that must be reformed:

> Pray bring to your Mind how often I desired you to consider, when you insisted on the Motive of *publick Good*, that the *Yahoos* were a Species of Animals utterly incapable of Amendment by Precepts or Examples: And so it has proved; for, instead of seeing a full Stop put to all Abuses and Corruptions, at least in this little Island, as I had Reason to expect: Behold, after above Six Months Warning, I cannot learn that my Book has produced one single Effect according to my Intentions. (Swift 1958: xxii)

The sharp point of Swift's wit as it is aimed at rationalist philosophies, which ask us to live our lives in accordance with reason, relates to the perceived difficulty of their application to people in general. That is, Swift interrogates the idea that the enlightenment that rational philosophy promises might only be attained by a few, and even then only after they have taken immense pains. A number of questions therefore arise. If it is so rare, can it be said to exist at all? Are those who claim to live under the enlightenment it brings only deluded? Or worse, does it lead them to become falsely superior and disdainful of their moral inferiors, so that in effect they themselves are immured in pride? That is, though they fail to see it, do they not remain among the filth of the world?

The idea of rarity is indeed affirmed by the famous final line of the *Ethics*, 'all things excellent are as difficult as they are rare' (E VP42S). This might be taken to imply that Spinoza's philosophy is not for the common people; that it might, rather, be only for the few, and this kind of view of 'rarity' is subjected to critique by Swift. A similar sentiment of rarity is also there at the beginning of the *Tractatus Theologico-Politicus* where Spinoza states that:

> I believe the main points are well enough known to philosophers [i.e. those capable of rational reasoning]. As for others, I am not particularly eager to recommend this treatise to them, for I have no reason to expect that it could please them in any way. I know how obstinately those prejudices stick in the mind that the heart has embraced in the form of piety. I know that it is as impossible to rid the common people of superstition as it is to rid them of fear. I know that the constancy of the common people is obstinacy, and that they are not governed by reason but swayed by impulse in approving or finding fault. I do not therefore invite the common people and those who are affected with the same feelings as they are [i.e. who think theologically], to read these things. I would wish them to ignore the book altogether rather than make a nuisance of themselves by interpreting it perversely, as they do with everything. (TTP Pref./G III 12)

The charge of elitism, however, can be answered. First, as the *Tractatus* and the *Ethics* demonstrate, Spinoza is highly conscious of the limitations placed upon people both by disposition and by education. One of the goals of the *Tractatus* is to question the manner in which people are taught by authority. He notes that while both the common people and theologians have attempted to align the Bible with their own 'false notions' and to use it to give authority to these notions, what is important in the Bible – its moral doctrines, such as that God is one and omnipotent, and that one should love one's neighbour as oneself (TTP ch. 7/G III 102) – is clear to all and is not the subject of disagreement among scholars: 'For the teachings of true piety are expressed in the most everyday language, since they are very common and extremely simple and easy to understand' (TTP ch. 7/G III 111). He further states that these truths and the word of God are written in people's hearts and that the light of reason is common to all. His goal, then, is to refute and set aside the curious interpretations attributed to the Bible and to affirm what is clear and distinct within it: its moral teaching. That is, it is to allow people to free themselves of the constraints of false reason. The comments directed to the 'common people' at the beginning of the *Tractatus* refer to those who are too bound by superstition to wish to attempt this, rather than to all people.

Secondly, Spinoza's circle looked for ways of educating the common people, including women who had been denied education, through dictionaries, encyclopedias and stories. The Spinozistic novel, as Israel calls it, sought to spread ideas by making them more accessible, through demonstration.

Swift makes us think about the potential limits of rational philosophy and the extent to which pure reason is compatible with the foibles of human nature. Yet it is clear that Spinoza himself was absolutely aware of human weakness, and directs Books 3 and 4 of the *Ethics* towards addressing these weaknesses. The idea of difficulty and rarity that I cite above from the end of the *Ethics*, rather than being read as elitist, might be understood merely to reflect the kinds of difficulties involved in realigning one's life to the dictates of reason; difficulties that pertain to all people, however ill- or well-educated.

Equally, it is clear from the *Tractatus* that Spinoza is aware of the potential of fiction as a way of cutting through and offering insight to beings such as ourselves, who, due to the conditions of our existence, are better suited to engaging with the imagination than directly with philosophical reason. He states:

> It . . . becomes clear why the prophets understood and taught almost everything in parables and allegorically, expressing all spiritual matters in corporeal language; for the latter are well suited to the nature of our imagination. (TTP ch. 1/G III 26)

This idea, that works of fiction might help to lead us to the truth, is further underlined by the use made of fiction by Spinoza's followers, who wrote works such as *Philopater* and translated works such as *Hayy Ibn Yaqzan*. It is further clear that the thinking through the imagination that is possible in fiction can help us to consider issues of philosophical and rational importance, such as the nature of relation and proportion in Spinoza's theory of knowledge.

Notes

1. This essay forms part of an Australian Research Council funded project with Moira Gatens (DP 170102206).
2. On Rochester, see Berman (1964: 359) and Ellenzweig (2005: 38). With regard to Alexander Pope, after the publication of *An Essay on Man* (a philosophical poem beloved by philosophers, including Voltaire and Kant), Pope was forced to strenuously deny that the system described in that poem, which seems to identify God with nature, was drawn from Spinoza or was Spinozist (Barnard 2009: 21, 312–13).
3. On Defoe, see Huebert (1948) and Novak (1961). On Dryden, see Hooker (1957: 187). Other connections have been traced between Spinoza and Henry Fielding (Smith 1961; 1962) and Laurence Sterne (Nowka 2009). For a comprehensive list of mentions of Spinoza in English, see Boucher (1999).
4. While the view of the Orient in Ibn Tufayl's reception is simplified and misrepresented, as Aravamudan has shown, it is apparent that Ibn Tufayl was used in two ways. First, the ideas he develops in his book directly inform the ideas of many European philosophers. Secondly, he offers an image of a world in which reasoning is tolerated, against a world, the Europe of the late seventeenth and early eighteenth centuries, where such freedom of reasoning might be condemned as heterodox.
5. That is, for 1, 2, and 3: $2 \times 3 = 6$; $6 \div 1 = 6$; therefore the fourth term is 6: i.e., 1 is to 2 as 3 is to 6.
6. I discuss this in detail elsewhere (Uhlmann 2011: 12–15).
7. These examples bring to mind Moses and Adam, both figures of importance to Spinoza. My thanks to Beth Lord for pointing this out.
8. Such a reading might involve comparing Spinoza's understanding of substance, Ibn Tufayl's understandings of God, and the paradoxical understandings of the One and the many set out in Plato's *Parmenides*, which was known to both Spinoza and Ibn Tufayl.
9. See, for example, chapter 6 of 'A Voyage to Brobdingnag', where he offers the king a detailed overview of the affairs and cultural organisation of contemporary England (Swift 1958: 100–1).
10. See Gardiner's argument, drawing on Zirker (1997), on the origin of the name Yahoo, and its relation to Jews and then to Christian believers (seen to be dupes following Tindall's criticism of those who believe in miracles). In sum, Yahoos are how a rationalist might see gullible Christians (Gardiner 2000: 235).

Notes on Contributors

Simon B. Duffy is Senior Lecturer at Yale-NUS College, Singapore. His research interests include early modern philosophy, modern and contemporary European philosophy, and the history and philosophy of science. He is the author of *Deleuze and the History of Mathematics: In Defense of the New* (2013) and *The Logic of Expression: Quality, Quantity and Intensity in Spinoza, Hegel and Deleuze* (2006), as well as numerous journal articles. He is editor of *Virtual Mathematics: The Logic of Difference* (2006), and co-editor of *Badiou and Philosophy* (2012). He is translator of Albert Lautman's *Mathematics, Ideas and the Physical Real* (2011). He has also translated a number of Gilles Deleuze's Seminars on Spinoza.

Hélène Frichot is an Associate Professor and Docent in the School of Architecture and the Built Environment, KTH, Stockholm. Between 2005 and 2014 she co-curated the Architecture+Philosophy public lecture series in Melbourne, Australia. In 2016 she co-convened the 13th international AHRA (Architectural Humanities Research Association) conference, which culminated in a book co-edited with Catharina Gabrielsson and Helen Runting called *Architecture and Feminisms: Ecologies, Economies, Technologies* (2017). She is currently on a Riksbankens Jubileumsfond funded sabbatical, working on a monograph entitled *Architectural Environment – Worlds, Things and Thinkables* (forthcoming 2018).

Gökhan Kodalak is an architectural designer, theorist, and historian. He is a PhD candidate in history of architecture at Cornell University. He is co-founder of ABOUTBLANK, an Istanbul-based design practice operating at the intersection of architecture, urbanism, and spatial research. His design work has been recognised with international awards, and exhibited in New York, Plovdiv, and Antalya, and he regularly presents his research at academic conferences and public lectures. He has published extensively on open-source architecture, Spinoza's affective aesthetics, the politics of the built environment and common space, Deleuze's diagrammatic approach to art and architecture, and metropolitan ecology. In his dissertation, he aims to unearth the peculiar potentials that underlie the missed encounter between Spinoza's philosophy and architecture.

Michael LeBuffe is Professor and Baier Chair of Early Modern Philosophy at the University of Otago. He is the author of *From Bondage to Freedom: Spinoza on Human Excellence* (2010) and *Spinoza on Reason* (2017). He has current research projects on Descartes, Hobbes, Spinoza, reason, virtue, and beauty.

Beth Lord is Reader in Philosophy at the University of Aberdeen. She was Principal Investigator on the AHRC *Equalities of Wellbeing* project which aimed to bring Spinoza's philosophy to bear on questions of architecture and housing design. She is the author of *Spinoza's Ethics: An Edinburgh Philosophical Guide* (2010) and *Kant and Spinozism: Transcendental Philosophy and Immanence from Jacobi to Deleuze* (2011), and editor of *Spinoza Beyond Philosophy* (2012). She is working on a book on Spinoza and equality.

Heidi M. Ravven is the Bates and Benjamin Professor of Classical and Religious Studies at Hamilton College. She is a Fellow in Neurophilosophy of the Integrative Neurosciences Research Program and a member of the Atrocity Prevention Study Group (Washington, DC), and the Virtues of Attention Project (New York University). She received a grant from the Ford Foundation in 2004 to write a book rethinking ethics: *The Self beyond Itself: An Alternative History of Ethics, the New Brain Sciences, and the Myth of Free Will* was published in 2013. She has published on many aspects of Spinoza's philosophy, focusing particularly on his relationships to affective neuroscience and the twelfth-century Jewish philosopher Maimonides.

Peg Rawes is Professor of Architecture and Philosophy and Director of the Masters in Architectural History at the Bartlett School of Architecture, UCL. Her research focuses on social and architectural histories of wellbeing, especially in contemporary housing, ecologies, and poetics. Her publications from the AHRC *Equalities of Wellbeing* project include the film *Equal by Design* (2016), and 'Housing Biopolitics and Care' in *Critical and Clinical Cartographies* (2017). Other recent publications include 'Planetary Aesthetics' in *Landscape and Agency* (2017) and the edited volumes *Poetic Biopolitics: Practices of Relation in Architecture and the Arts* (2016), and *Relational Architectural Ecologies* (2013).

Anthony Uhlmann is Director of the Writing and Society Research Centre at Western Sydney University. He is the author of three monographs: *Beckett and Poststructuralism* (Cambridge University Press, 1999), *Samuel Beckett and the Philosophical Image* (Cambridge University Press, 2006), and *Thinking in Literature: Joyce, Woolf, Nabokov* (Bloomsbury, 2011). He has recently completed a new work on J. M. Coetzee and is currently working on a project on Spinoza and literature.

Valtteri Viljanen is an Academy of Finland Research Fellow at the University of Turku. He is the author of *Spinoza's Geometry of Power* (Cambridge University Press, 2011) and numerous articles on Spinoza.

Stefan White is Professor of Architecture at Manchester School of Architecture. He researches and practises the architecture and urbanism of social and environmental sustainability. He is co-director of the PHASE (Place-Health Architecture Space Environment) research and delivery group at the Manchester School of Architecture, working collaboratively to understand and create healthier places with the UK government, regional health and care providers, city councils, registered housing providers, and local communities.

Timothy Yenter is Assistant Professor of Philosophy at the University of Mississippi. His research focuses on metaphysics and methodology in the seventeenth and eighteenth centuries, with special attention to the contested standards and uses of demonstrations in philosophy. His forthcoming work includes essays on Newton's impact on Scottish metaphysics in the eighteenth century and the contemporary relevance of Hume's writings on religion.

Bibliography

Adler, Jacob (2014), 'Mortality of the Soul from Alexander of Aphrodisias to Spinoza', in Steven Nadler (ed.), *Spinoza and Medieval Jewish Philosophy*, Cambridge: Cambridge University Press, pp. 13–35.
Althusser, Louis (2006 [1982]), 'The Underground Current of the Materialism of the Encounter', in *Philosophy of the Encounter: Later Writings, 1978–87*, ed. F. Matheron and O. Corpet, London: Verso, pp. 163–207.
Anwan, N., T. Schneider, and J. Till (2011), *Spatial Agency: Other Ways of Doing Architecture*, London: Routledge.
Aravamudan, Srinivas (2011), *Enlightenment Orientalism: Resisting the Rise of the Novel*, Chicago: University of Chicago Press.
Aristotle (1970), *Nicomachean Ethics*, in *The Basic Works of Aristotle*, ed. Richard McKeon, New York: Random House, pp. 935–1126.
Armstrong, Aurelia (2009), 'Autonomy and the Relational Individual: Spinoza and Feminism', in Moira Gatens (ed.), *Feminist Interpretations of Benedict Spinoza*, University Park, PA: Pennsylvania State University Press, pp. 43–64.
Bakhtin, Mikhail (2009), *Rabelais and his World*, trans. Hélène Iswolsky, Bloomington, IN: Indiana University Press.
Balibar, Etienne (1997), 'Spinoza: From Individuality to Transindividuality', *Mededelingen vanwege het Spinozahuis* 71, Eburon Delft.
Balibar, Etienne (2008 [1985]), *Spinoza and Politics*, trans. Peter Snowdon, London: Verso.
Barbaras, Françoise (2007), *Spinoza: la science mathematique du salut*, Paris: CNRS.
Barnard, John (2009), *Alexander Pope: The Critical Heritage*, London: Routledge.
Bateson, Gregory (1987 [1972]), *Steps to an Ecology of Mind*, Northvale, NJ: Jason Aronson.
Bell, David (1984), *Spinoza in Germany from 1670 to the Age of Goethe*, London: Institute of Germanic Studies.
Bennett, Jane (2010), *Vibrant Matter: A Political Ecology of Things*, Durham, NC: Duke University Press.
Bennett, Jonathan (1984), *A Study of Spinoza's Ethics*, Cambridge: Cambridge University Press.
Bennett, Jonathan (2001), *Learning from Six Philosophers: Descartes, Spinoza, Leibniz, Locke, Berkeley, Hume*, vols 1 and 2, New York: Oxford University Press.

Berrien, Kenneth F. (1968), *General and Social Systems*, New Brunswick, NJ: Rutgers University Press.
Berman, Ronald (1964), 'Rochester and the Defeat of the Senses', *The Kenyon Review* 26(2): 354–68.
Blair, Ann (1997), *The Theatre of Nature: Jean Bodin and Renaissance Science*, Princeton: Princeton University Press.
Boucher, Wayne (1999), *Spinoza in English: A Bibliography*, 2nd edn, London: Thoemmes.
Bowie, D. (2017), *Radical Solutions to the Housing Supply Crisis*, Bristol: Policy Press.
Brünner, Margit (2011), 'Becoming Joy: Experimental Construction of Atmospheres', PhD thesis, University of South Australia.
Brünner, Margit (2015), *Constructing Atmospheres: Test-Sites for an Aesthetics of Joy*, Baunach: Art Architecture Design Research, Spurbuchverlag.
Buursma, G. (2011), *The Hague: An Architectural Guide*, Rotterdam: 010 Publishers.
Carmona, M. (2010), *Space Standards: The Benefits*, Report for CABE, London: UCL.
Chalmers, David (2006), 'Strong and Weak Emergence', in P. Clayton and P. Davies (eds), *The Re-emergence of Emergence*, Oxford: Oxford University Press, pp. 244–57.
Clarke, Samuel (1738), *Works*, ed. Benjamin Hoadly, 4 vols, London: John and Paul Knapton.
Colerus, John (1706), *The Life of Benedict De Spinosa*, London: D.L.
Colie, Rosalie L. (1959), 'Spinoza and the Early English Deists', *Journal of the History of Ideas* 20(1): 23–46.
Colie, Rosalie L. (1963), 'Spinoza in England', *Proceedings of the American Philosophical Society* 107(3): 183–219.
Conway, Anne (1996 [1690]), *The Principles of the Most Ancient and Modern Philosophy*, ed. Allison P. Coudert and Taylor Corse, New York: Cambridge University Press.
Damasio, Antonio (2003), *Looking for Spinoza: Joy, Sorrow, and the Feeling Brain*, Orlando: Harvest.
Defoe, Daniel (2003), *Robinson Crusoe*, ed. John Richetti, London: Penguin.
DeLanda, Manuel (2002), *Intensive Science and Virtual Philosophy*, London: Continuum.
Deleuze, Gilles (1978), 'Spinoza. Cours Vincennes', Paris, Université de Paris VIII Vincennes, https://www.webdeleuze.com/groupes/2 (accessed 26 April 2017).
Deleuze, Gilles (1988a), *Bergsonism*, trans. Hugh Tomlinson and Barbara Habberjam, New York: Zone Books.
Deleuze, Gilles (1988b), *Foucault*, trans. and ed. Sean Hand, London: Athlone.
Deleuze, Gilles (1988c), *Spinoza: Practical Philosophy*, trans. Robert Hurley, San Francisco: City Lights Books.

Deleuze, Gilles (1991), 'A Philosophical Concept...', in Eduardo Cadava et al. (eds), *Who Comes After the Subject?*, London: Routledge, pp. 94–5.
Deleuze, Gilles (1992 [1990]), *Expressionism in Philosophy: Spinoza*, trans. M. Joughin, New York: Zone Books.
Deleuze, Gilles (1994), *Difference and Repetition*, trans. Paul Patton, London: Continuum.
Deleuze, Gilles (1998), *Essays Critical and Clinical*, trans. Daniel W. Smith and Michael A. Greco, London: Verso.
Deleuze, Gilles (2001), 'Immanence: A Life...', in *Pure Immanence: Essays on a Life*, trans. Anne Boyman, New York: Zone Books.
Deleuze, Gilles (2003), 'The Three Kinds of Knowledge', *Pli: The Warwick Journal of Philosophy* 14: 1–20.
Deleuze, Gilles, and Félix Guattari (1994), *What is Philosophy?*, trans. Hugh Tomlinson and Graham Burchill, New York: Columbia University Press.
Deleuze, Gilles, and Félix Guattari (2005 [1980]), *A Thousand Plateaus: Capitalism and Schizophrenia*, trans. Brian Massumi, Minneapolis: University of Minnesota Press.
Della Rocca, Michael (1994), 'Egoism and the Imitation of the Affects in Spinoza', in *Ethica IV: Spinoza on Reason and the 'Free Man'*, papers presented at the Fourth Jerusalem Conference, New York: Little Room Press, pp. 123–47.
Della Rocca, Michael (1996), *Representation and the Mind–Body Problem in Spinoza*, New York: Oxford University Press.
Della Rocca, Michael (2012), 'Rationalism, Idealism, Monism, and Beyond', in E. Förster and Y. Melamed (eds), *Spinoza and German Idealism*, Cambridge: Cambridge University Press, pp. 7–26.
Derrida, Jacques (1998), 'Point de folie – Maintenant l'architecture', in M. Hays (ed.), *Architecture Theory Since 1968*, Cambridge, MA: MIT Press, pp. 570–81.
Descartes, René (1985), *The Philosophical Writings of Descartes*, vol. I, trans. John Cottingham, Robert Stoothoff, and Dugald Murdoch, Cambridge: Cambridge University Press.
Descartes, René (1996), *Meditations on First Philosophy*, trans. and ed. John Cottingham, Cambridge: Cambridge University Press.
Drury, A. (2008), 'Parker Morris – Holy Grail or Wholly Misguided?', *Town and Country Planning* 77(10): 403–5.
Duffy, Simon (2006), *The Logic of Expression: Quality, Quantity and Intensity in Spinoza, Hegel and Deleuze*, Aldershot: Ashgate.
Duns Scotus (1987), *Philosophical Writings*, ed. A. Wolter, Indianapolis: Hackett.
Durie, Robin (2002a), 'Immanence and Difference: Towards a Relational Ontology', *Southern Journal of Philosophy* 60: 1–30.
Durie, Robin (2002b), 'Creativity and Life', *The Review of Metaphysics* 56(2): 357–83.
DWELL (Designing for Wellbeing in Environments for Later Life) (2016), http://dwell.group.shef.ac.uk/ (accessed 3 January 2017).
Edwards, Jonathan (1980 [1714]), 'The Mind', in *The Works of Jonathan Edwards*,

vol. 6: The Scientific and Philosophical Writings, ed. Wallace E. Anderson, New Haven, CT: Yale University Press, pp. 332–86.
Edwards, Jonathan (1994 [1722]), 'The Miscellanies', in *The Works of Jonathan Edwards, vol. 13*, ed. Thomas Schafer, New Haven, CT: Yale University Press.
Eisenman, P. (1995), 'Eisenman (and Company) Respond: The Politics of Formalism', *Progressive Architecture* 76(2): 88–91.
Eliot, John (1678), *The Harmony of the Gospels, in the Holy History of the Humiliation and Sufferings of Jesus Christ, from His Incarnation to His Death and Burial. / Published by John Eliot, Teacher of the Church in Roxbury.; [Two Lines from Acts]*, Boston: Printed by John Foster, http://name.umdl.umich.edu/N00187.0001.001 (accessed 13 October 2017).
Ellenzweig, Sarah (2005), 'The Faith of Unbelief: Rochester's "Satyre", Deism, and Religious Freethinking in Seventeenth-Century England', *Journal of British Studies* 44(1): 27–45.
Ely, A. (2015), 'The Case for Space', paper given at the Equalities of Wellbeing conference, UCL, April 2015, http://www.equalitiesofwellbeing.co.uk/publications-from-equalitiesof-wellbeing-housing-workshop/ (accessed 14 July 2015).
Euclid (1908), *The Thirteen Books of the Elements*, 2nd edn, trans. T. L. Heath, Cambridge: Cambridge University Press.
Evans, Robin (1995), *The Projective Cast: Architecture and its Three Geometries*, Cambridge, MA: MIT Press.
Evans, Robin (1997), *Translations between Drawing and Building and Other Essays*, London: Architectural Association Publications.
Frichot, Hélène (2009), 'On Finding Oneself Spinozist: Refuge, Beatitude and the Any-Space-Whatever', in Charles J. Stivale, Eugene W. Holland, and Daniel W. Smith (eds), *Gilles Deleuze: Image and Text*, London: Continuum, pp. 247–63.
Frichot, Hélène (2011), 'Drawing, Thinking, Doing: From Diagram Work to the Superfold', *ACCESS: Critical Perspectives on Communication, Cultural & Policy Studies* 30(1): 1–10.
Gabbey, Alan (1996), 'Spinoza's Natural Science and Methodology', in Don Garrett (ed.), *The Cambridge Companion to Spinoza*, Cambridge: Cambridge University Press, pp. 142–91.
Gardiner, Anne B. (2000), '"Be ye as the horse!" Swift, Spinoza and the Society of Virtuous Atheists', *Studies in Philology* 97: 229–53.
Garrett, Don (1991), 'Spinoza's Necessitarianism', in Yirmiyahu Yovel (ed.), *God and Nature: Spinoza's Metaphysics*, Leiden: Brill, pp. 191–218.
Garrett, Don (1994), 'Spinoza's Theory of Metaphysical Individuation', in J. Gracia and K. Barber (eds), *Individuation in Early Modern Philosophy*, Albany: State University of New York Press, pp. 73–97.
Garrett, Don (2009), 'Spinoza on the Essence of the Human Body and the Part of the Mind that is Eternal', in Olli Koistinen (ed.), *The Cambridge Companion to Spinoza's Ethics*, Cambridge: Cambridge University Press, pp. 284–302.
Garvie, D., and Shelter (2015), 'Little Boxes, Fewer Homes – Why Housing Space Standards Will Get More Homes Built', paper given at the Equalities

of Wellbeing conference, April 2015, http://www.equalitiesofwellbeing.co.uk/publications-from-equalities-ofwellbeing-housing-workshop/ (accessed 14 July 2015).

Gatens, Moira (2013), 'Cloud-Borne Angels, Prophets, and the Old Woman's Flower-pot: Reading George Eliot's Realism alongside Spinoza's "beings of the imagination"', *Australian Literary Studies* 28(3): 1–14.

Gatens, Moira, and Genevieve Lloyd (1999), *Collective Imaginings: Spinoza, Past and Present*, London: Routledge.

Ghirardo, D. (1994), 'Eisenman's Bogus Avant-Garde', *Progressive Architecture* 72: 70–3.

Godfrey-Smith, Peter (2017), *Other Minds: The Octopus and the Evolution of Intelligent Life*, London: William Collins.

Goose, N., and H. Looijesteijn (2012), 'Almshouses in England and the Dutch Republic *circa* 1350–1800: A Comparative Perspective', *Journal of Social History* 45(4): 1049–73.

Gregg, M., and G. Seigworth (eds) (2010), *The Affect Theory Reader*, Durham, NC: Duke University Press.

Grey, John (2013), '"Use Them At Our Pleasure": Spinoza on Animal Ethics', *History of Philosophy Quarterly* 30(4): 367–88.

Grosz, Elizabeth (2001), 'The Natural in Architecture and Culture', in *Architecture from the Outside*, Cambridge, MA: MIT Press, pp. 91–109.

Guattari, Félix (1995), *Chaosmosis: An Ethico-Aesthetic Paradigm*, trans. Paul Bains and Julian Pefanis, Sydney: Power Publications.

Gueroult, Martial (1968), *Spinoza I. Dieu (Éthique, I)*, Paris: Aubier-Montaigne.

Gueroult, Martial (1974), *Spinoza II. L'Âme (Éthique, II)*, Paris: Aubier-Montaigne.

Hampshire, Stuart (1996), *Ethics: Benedict de Spinoza*, London: Penguin.

Haraway, Donna (1988), 'Situated Knowledges: The Science Question in Feminism and the Privilege of Partial Perspective', *Feminist Studies* 14(3): 575–99.

Harker, L. (2006), *Chance of a Life Time: The Impact of Bad Housing on Children's Lives*, London: Shelter.

HATC (Housing Association Training and Consultancy) (2006), *Housing Space Standards*, London: Greater London Authority.

Hawking, Stephen (1998), *A Brief History of Time*, New York: Bantam.

Heidegger, Martin (1995), *The Fundamental Concepts of Metaphysics: World, Finitude, Solitude*, trans. William McNeill and Nicholas Walker, Indianapolis: Indiana University Press.

Heydon, John (1662), *The Harmony of the World: Being a Discourse of God, Heaven, Angels, Stars, Planets, Earth, the Miraculous Descentions and Ascentions of Spirits: With the Nature and Harmony of Mans Body, the Art of Preparing Rosie Crucian Medicines to Cure All Diseases: Their Rules to Raise Bodies Decayed, Which Are Verified by a Practical Examination of Principles in the Great World: Whereunto Is Added, the State of the New Jerusalem, Grounded upon the Knowledge of Nature, Light of Reason, Phylosophy and Divinity: All Fitted to the Understanding, Use*

and Profit of Wisdomes Children, and Communicated to the Sons of Art, London: Robert Horn.
Hobbes, Thomas (1998 [1651]), *Leviathan*, Oxford: Oxford University Press.
Hobbes, Thomas (2003 [1642]), *On the Citizen*, Cambridge: Cambridge University Press.
Hooker, Edward N. (1957), 'Dryden and the Atoms of Epicurus', *ELH* 24(3): 177–90.
Hooton, William R., III (1999), 'Wordsworth, Coleridge and the Politics of Pantheism', *Coleridge Bulletin: The Journal of the Friends of Coleridge* 14: 60–72.
Hübner, Karolina (2016), 'Spinoza on Essences, Universals, and Beings of Reason', *Pacific Philosophical Quarterly* 97(1): 58–88.
Huebert, Lee (1948), *Defoe and Deism*, Lincoln: University of Nebraska Press.
Huenemann, Charlie (1999), 'The Necessity of Finite Modes and Geometrical Containment in Spinoza's Metaphysics', in R. J. Gennaro and C. Huenemann (eds), *New Essays on the Rationalists*, New York: Oxford University Press, pp. 224–40.
Hutcheson, Francis (2006 [1742]), *Logic, Metaphysics, and the Natural Sociability of Mankind*, ed. James Moore and Michael Silverthorne, trans. Michael Silverthorne, Natural Law and Enlightenment Classics, Indianapolis: Liberty Fund.
Hutcheson, Francis (2008 [1738]), *An Inquiry into the Original of Our Ideas of Beauty and Virtue*, ed. Wolfgang Leidhold, Indianapolis: Liberty Fund.
Ibn Arabi, Muhyiddin (1980), *The Bezels of Wisdom*, New York: Paulist Press.
Ibn Tufayl (2003), *Ibn Hayy Yaqzan: A Philosophical Tale*, trans. Lenn Evan Goodman, Chicago: University of Chicago Press.
Israel, Jonathan I. (2001), *Radical Enlightenment: Philosophy and the Making of Modernity 1650–1750*, Oxford: Oxford University Press.
Israel, Jonathan I. (2006), *Enlightenment Contested: Philosophy, Modernity, and the Emancipation of Man 1670–1752*, Oxford: Oxford University Press.
Ivry, Alfred (2009), 'Maimonides' Psychology', in Idit Dobbs-Weinstein et al. (eds), *Maimonides and his Heritage*, Albany: SUNY Press, pp. 51–60.
James, Susan (1996), 'Power and Difference: Spinoza's Conception of Freedom', *Journal of Political Philosophy* 4(3): 207–28.
James, Susan (1997), *Passion and Action: The Emotions in Seventeenth Century Philosophy*, Oxford: Oxford University Press.
Jaquet, Chantal (2004), *L'unité du corps et de l'esprit: Affects, actions et passions chez Spinoza*, Paris: Presses Universitaires de France.
Jarrett, Charles (1990), 'The Development of Spinoza's Conception of Immortality', in F. Mignini (ed.), *Dio, l'uomo, la liberta: studi sul Breve Trattato di Spinoza. Actes du colloque de l'Aquila, 20–23 October, 1987*, L'Aquila-Rome: L. U. Japadre, pp. 147–88.
Jarrett, Charles (2009), 'Spinoza on Necessity', in Olli Koistinen (ed.), *The Cambridge Companion to Spinoza's Ethics*, Cambridge: Cambridge University Press, pp. 118–39.
Jones, R. (2016), 'Fancy Living in a Historic City Centre Pad for about £120

a Week?', *The Guardian*, 17 September, https://www.theguardian.com/money/2016/sep/17/historic-city-centre-pad-almshouses-tenants (accessed 5 May 2017).
Joseph Rowntree Foundation (2015), Housing and Poverty Blog, http://www.jrf.org.uk/topic/housing-and-poverty (accessed 14 July 2015).
Joyce, David E. (trans. and ed.) (1998), *Euclid's Elements*, http://aleph0.clarku.edu/~djoyce/elements/elements.html (accessed 1 August 2017).
Koistinen, Olli (1998), 'On the Consistency of Spinoza's Modal Theory', *The Southern Journal of Philosophy* 36(1): 61–80.
Kostof, Spiro (1977), 'The Architect in the Middle Ages, East and West', in *The Architect: Chapters in the History of the Profession*, Oxford: Oxford University Press, pp. 67–93.
Kwinter, Sanford (2002), *Architectures of Time: Towards a Theory of the Event in Modernist Culture*, Cambridge, MA: MIT Press.
Lærke, Mogens (2008), *Leibniz lecteur de Spinoza: la genèse d'une opposition complexe*, Paris: Honoré Champion.
Lamont, Thomas (2002), 'Mutual Abuse: The Meeting of Robinson Crusoe and Hayy Ibn Yaqzân', *Edebiyat: Journal of M.E. Literatures* 13(2): 169–76.
LeBuffe, Michael (2010), 'Change and the Eternal Part of the Mind in Spinoza', *Pacific Philosophical Quarterly* 91: 369–84.
LeBuffe, Michael (2017), *Spinoza on Reason*, New York: Oxford University Press.
Le Clerc, John (1701), *The Harmony of the Evangelists, with a Paraphrase, with Dissertations*, London: Sam. Buckley at the Dolphin in St Paul's Church-yard.
Lefebvre, Henri (1991), *The Production of Space*, Cambridge: Blackwell.
Leibniz, Gottfried Wilhelm (1989), *Philosophical Essays*, trans. Roger Ariew and Daniel Garber, Indianapolis: Hackett.
Leibniz, Gottfried Wilhelm (1998), *Philosophical Texts*, trans. and ed. R. S. Woolhouse and R. Francks, Oxford: Oxford University Press.
Leibniz, Gottfried Wilhelm (2004), *Confessio Philosophi: Papers Concerning the Problem of Evil, 1671–1678*, trans. Robert C. Sleigh, The Yale Leibniz, New Haven, CT: Yale University Press.
Lloyd, Genevieve (1994), *Part of Nature: Self-Knowledge in Spinoza's Ethics*, Ithaca: Cornell University Press.
Locke, John (1824 [1690]), 'Some Thoughts Concerning Education', in *The Works of John Locke*, 12th edn, London: C. and J. Rivington.
Lord, Beth (2014), 'Spinoza, Equality, and Hierarchy', *History of Philosophy Quarterly* 31(1): 59–77.
Lord, Beth (2016), 'The Concept of Equality in Spinoza's *Theological-Political Treatise*', *Epoche: A Journal for the History of Philosophy* 20(2): 367–86.
Lord, Beth (ed.) (2012), *Spinoza Beyond Philosophy*, Edinburgh: Edinburgh University Press.
Lovejoy, Arthur (2001), *The Great Chain of Being*, Cambridge, MA: Harvard University Press.
Macherey, Pierre (1995), *Introduction à L'Ethique de Spinoza, la Troisième Partie*, Paris: Presses Universitaires de France.

Macherey, Pierre (2011), *Hegel or Spinoza*, trans. Susan M. Ruddick, Minneapolis: University of Minnesota Press.
McDowell, Paula (1998), *The Women of Grub Street: Press, Politics, and Gender in the London Literary Marketplace 1678–1730*, Oxford: Clarendon Press.
Macknight, James (1763), *A Harmony of the Four Gospels: In Which the Natural Order of Each Is Preserved: With a Paraphrase and Notes*, 2nd edn, London: William Strahan et al.
Malamud, Bernard (1966), *The Fixer*, New York: Farrar, Straus, and Giroux.
Malebranche, Nicholas (1700), *The Search after Truth*, trans. T. Taylor, 2nd edn, London: W. Bowyer.
Manning, Richard (2012), 'Spinoza's Physical Theory', in *The Stanford Encyclopedia of Philosophy*, ed. Edward N. Zalta, http://plato.stanford.edu/archives/spr2012/entries/spinoza-physics/ (accessed 1 August 2017).
Marshall, Eugene (2008), 'Adequacy and Innateness in Spinoza', in Daniel Garber and Steven Nadler (eds), *Oxford Studies in Early Modern Philosophy*, Oxford: Clarendon Press, vol. 4, pp. 51–88.
Marshall, Eugene (2013), *The Spiritual Automaton: Spinoza's Science of the Mind*, Oxford: Oxford University Press.
Martin, Christopher P. (2008), 'The Framework of Essences in Spinoza's *Ethics*', *British Journal for the History of Philosophy* 16(3): 489–509.
Massumi, Brian (2002), *Parables for the Virtual: Movement, Affect, Sensation*, Durham, NC: Duke University Press.
Matheron, Alexandre (1969), *Individu et communauté chez Spinoza*, Paris: Editions de Minuit.
Matheron, Alexandre (1986), 'Spinoza and Euclidean Arithmetic: The Example of the Fourth Proportional', in Marjorie Grene and Debra Nails (eds), *Spinoza and the Sciences*, Dordrecht: D. Reidel, pp. 125–50.
Maturana, Humberto, and Francisco Varela (1980), *Autopoiesis and Cognition: The Realization of the Living*, Dordrecht: D. Reidel.
Mayor of London (2010a), *Housing Design Standards*, Evidence Summary, London: Mayor of London.
Mayor of London (2010b), *Design for London, London Housing Design Guide* (Interim Edition), London: Mayor of London.
Melamed, Yitzhak Y. (2013), *Spinoza's Metaphysics: Substance and Thought*, Oxford: Oxford University Press.
Minton, A. (2017), *Big Capital: Who is London For?* London: Penguin.
Mitchell, William (2004), *Me++: The Cyborg Self and the Networked City*, Cambridge, MA: MIT Press.
Montag, Warren (1994), *The Unthinkable Swift: Spontaneous Philosophy in the Church of England Man*, London: Verso.
Montag, Warren (1999), *Bodies, Masses, Power: Spinoza and his Contemporaries*, London: Verso.
Montag, Warren (2005), 'Who's Afraid of the Multitude? Between the Individual and the State', *The South Atlantic Quarterly* 104(4): 655–73.

Montag, Warren, and Ted Stolze (eds) (1997), *The New Spinoza*, Minneapolis: University of Minnesota Press.

More, Henry (1653), *Conjectura Cabbalistica Or, a Conjectural Essay of Interpreting the Minde of Moses, according to a Threefold Cabbala: Viz. Literal, Philosophical, Mystical, Or, Divinely Moral*, London: James Flesher.

More, Henry (1668), *The Two Last Dialogues Treating of the Kingdome of God within Us and without Us, and of His Special Providence through Christ over His Church from the Beginning to the End of All Things: Whereunto Is Annexed a Brief Discourse of the True Grounds of the Certainty of Faith in Points of Religion, Together with Some Few Plain Songs of Divine Hymns on the Chief Holy-Days of the Year*, London: J. Flesher.

More, Henry (1681), *A Plain and Continued Exposition of the Several Prophecies or Divine Visions of the Prophet Daniel Which Have or May Concern the People of God, Whether Jew or Christian: Whereunto Is Annexed a Threefold Appendage Touching Three Main Points, the First Relating to Daniel, the Other Two to the Apocalypse*, London: M.F.

More, Henry (1692), *Discourses on Several Texts of Scripture*, London: J.R.

National Centre for Social Research and Shelter (2013), *People Living in Bad Housing – Numbers and Health Impacts*, London: Shelter.

Negri, Antonio (1991 [1981]), *The Savage Anomaly: The Power of Spinoza's Metaphysics and Politics*, Minneapolis: University of Minnesota Press.

Newlands, Samuel (2010), 'Another Kind of Spinozistic Monism', *Nous* 44(3): 469–502.

Novak, Maximillian E. (1961), 'Robinson Crusoe's Fear and the Search for Natural Man', *Modern Philology* 58(4): 238–45.

Nowka, Scott (2009), 'Talking Coins and Thinking Smoke-Jacks: Satirizing Materialism in Gildon and Sterne', *Eighteenth Century Fiction* 22(2): 195–222.

Numbeo (2017), 'Cost of Living in the Hague', https://www.numbeo.com/cost-of-living/in/The-Hague (accessed 5 January 2017).

Park, A., F. Ziegler, and S. Wigglesworth (2016), *Designing with Downsizers*, Report for DWELL, Sheffield: University of Sheffield.

Park, J. (2017), *One Hundred Years of Housing Space Standards. What Now?*, Housing Space Standards, http://www.housingspacestandards.co.uk/ (accessed 1 May 2017).

Parker Morris Report (1961), *Homes for Today and Tomorrow*, London: Ministry of Housing and Local Government.

Pelbart, Peter Pal (2015), *Cartography of Exhaustion: Nihilism Inside Out*, Minneapolis: Univocal Publishing.

Pepper, S. (2015), 'Three Ages of Post-war Housing', Housing and Design seminar, London, 30 January 2015, http://www.equalitiesofwellbeing.co.uk/podcasts-from-housing-and-design-seminar/podcast (accessed 14 July 2015).

Peter Barber Architects (2017), http://www.peterbarberarchitects.com/holmes-road-studios (accessed 5 January 2017).

Peterman, Alison (2014), 'Spinoza on Physical Science', *Philosophy Compass* 9: 214–23.

Pilkington, Matthew (1747), *The Evangelical History and Harmony*, London: William Bowyer.
Plato (2008), *Timaeus and Critias*, trans. Robin Waterfield, with an introduction and notes by Andrew Gregory, Oxford: Oxford University Press.
Popkin, Richard H. (2003), *The History of Scepticism: From Savonarola to Bayle*, London: Oxford University Press.
Price, Cedric, and Joan Littlewood (1968), 'The Fun Palace', *The Drama Review: TDR, Architecture/Environment* 12(3): 127–34.
Protevi, John (2009), *Political Affect: Connecting the Social and the Somatic*, Minneapolis: University of Minnesota Press.
Ramond, Charles (1995), *Qualité et quantité dans la philosophie de Spinoza*, Paris: Presses Universitaires de France.
Ravven, Heidi M. (1989), 'Notes on Spinoza's Critique of Aristotle's Ethics: From Teleology to Process Theory', *Philosophy and Theology* IV(1): 3–32.
Ravven, Heidi M. (1990), 'Spinoza's Materialist Ethics: The Education of Desire', *International Studies in Philosophy* XXII(3): 59–78.
Ravven, Heidi M. (2003), 'Spinoza's Anticipation of Contemporary Affective Neuroscience', *Consciousness & Emotion* 4(2): 257–90.
Ravven, Heidi M. (2004), 'Spinoza's Systems Theory of Ethics', in George E. Lasker, Iva Smith, and Wendell Wallach (eds), *Cognitive, Emotive, and Ethical Aspects of Decision Making in Humans and in Artificial Intelligence*, vol. III, Windsor, Ont.: The International Institute for Advanced Studies in Systems Research & Cybernetics Press, pp. 99–104.
Ravven, Heidi M. (2005), 'What Can Spinoza Teach Us Today about Naturalizing Ethics? Provincializing Philosophical Ethics and Freedom without Free Will', in George E. Lasker, Iva Smith, and Wendell Wallach (eds), *Cognitive, Emotive, and Ethical Aspects of Decision Making in Humans and in Artificial Intelligence*, vol. IV, Windsor, Ont.: The International Institute for Advanced Studies in Systems Research & Cybernetics Press, pp. 103–8.
Ravven, Heidi M. (2012), 'The Self Beyond Itself: Further Reflections on Spinoza's Systems Theory of Ethics', in Nagib Callaos et al. (eds), *Proceedings of the 3rd International Multi-Conference on Complexity, Informatics and Cybernetics*, Winter Garden, FL: IIIS Press, pp. 133–8.
Ravven, Heidi M. (2013). *The Self Beyond Itself: An Alternative History of Ethics, the New Brain Sciences, and the Myth of Free Will*, New York: The New Press.
Rawes, P., and B. Lord (2016), *Equal by Design*, film, directed by Adam Low, Lone Star Productions, www.equalbydesign.co.uk (accessed 15 December 2016).
Reynolds, L., and N. Robinson (2005), *Full House? How Overcrowded Housing Affects Families*, London: Shelter.
RIBA (Royal Institute of British Architects) (2011), *The Case for Space: The Size of England's New Homes*, London: Royal Institute of British Architects.
Rice, Lee C. (1990), 'Individual and Community in Spinoza's Social Psychology', in Edwin Curley and Pierre-Francois Moreau (eds), *Spinoza: Issues and Directions*, Leiden: Brill.
Roys, M., M. Davidson, S. Nicol, D. Ormandy, and P. Ambrose (2010), *The Real*

Cost of Poor Housing, Building Research Establishment Trust Report FB23, Watford: BRE Press.

Ryan-Collins, J., T. Lloyd, and L. Macfarlane (2017), *Rethinking the Economics of Land and Housing*, London: Zed Books.

Schmidt, Andreas (2009), 'Substance Monism and Identity Theory in Spinoza', in Olli Koistinen (ed.), *The Cambridge Companion to Spinoza's Ethics*, Cambridge: Cambridge University Press, pp. 79–98.

Shein, Noa (2015), 'Causation and Determinate Existence of Finite Modes in Spinoza', *Archiv für Geschichte der Philosophie* 97: 334–57.

Shelter (2013a), *Solutions to the Housing Shortage*, London: Shelter.

Shelter (2013b), *Little Boxes, Fewer Homes: Setting Housing Space Standards Will Get More Homes Built*, London: Shelter.

Shelter and KPMG (2013), *Homes for the Next Generation*, London: Shelter.

Smith, Leroy W. (1961), 'Fielding and Mandeville: The "War Against Virtue"', *Criticism* 3(1): 7–15.

Smith, Leroy W. (1962), 'Fielding and "Mr. Bayle's" Dictionary', *Texas Studies in Literature and Language* 4(1): 16–20.

Spinoza, Baruch [Benedict de] (1967), *Spinoza's Ethics and on the Correction of the Understanding*, London and New York: Everyman's Library.

Spinoza, Baruch (1985), *The Collected Works of Spinoza*, vol. I, trans. and ed. Edwin Curley, Princeton: Princeton University Press.

Spinoza, Baruch (1992), *Ethics, Treatise on the Emendation of the Intellect and Selected Letters*, trans. S. Shirley, ed. S. Feldman, Indianapolis: Hackett.

Spinoza, Baruch (1996), *Ethics*, trans. E. Curley, London: Penguin.

Spinoza, Baruch (2002), *Spinoza: Complete Works*, ed. M. Morgan, trans. S. Shirley, Indianapolis: Hackett.

Spinoza, Baruch (2007), *Theological-Political Treatise*, ed. Jonathan Israel, trans. Michael Silverthorne and Jonathan Israel, Cambridge: Cambridge University Press.

Steenbakkers, Piet (1994), *Spinoza's Ethica from Manuscript to Print: Studies on Text, Form and Related Topics*, Amsterdam: Van Gorcum.

Steenbakkers, Piet (2009), 'The Geometrical Order in the *Ethics*', in Olli Koistinen (ed.), *The Cambridge Companion to Spinoza's Ethics*, Cambridge: Cambridge University Press, pp. 42–55.

Stengers, Isabelle (2010), *Cosmopolitics I*, trans. Robert Bononno, Minneapolis: University of Minnesota Press.

Survey of London, The (1994), 'Public Housing in Poplar: The 1940s to the Early 1990s', in *Survey of London: Volumes 43 and 44, Poplar, Blackwall and Isle of Dogs*, ed. Hermione Hobhouse, London: London City Council, pp. 37–54, http://www.british-history.ac.uk/survey-london/vols43-4/ (accessed 1 May 2015).

Swenarton, M. (1981), *Homes Fit for Heroes: The Politics and Architecture of Early State Housing in Britain*, London: Ashgate.

Swift, Jonathan (1861), *The Works of Jonathan Swift ...: With Cop'ous Notes and Additions*, vol. 5, Ann Arbor, MI: University of Michigan Library.

Swift, Jonathan (1958), *Gulliver's Travels and Other Writings*, ed. Ricardo Quintana, New York: Modern Library.
Teyssot, Georges (1994), 'The Mutant Body of Architecture', in *Flesh: Architectural Probes*, New York: Princeton Architectural Press.
Thompson, Evan (2009), 'Life and Mind: From Autopoiesis to Neurophenomenology', in B. Clarke and M. Hansen (eds), *Emergence and Embodiment*, Durham, NC: Duke University Press, pp. 77–94.
Thrift, Nigel (2007), *Non-Representational Theory: Space, Politics, Affect*, London: Routledge.
Uexküll, Jakob von (2010), *A Foray into the Worlds of Animals and Humans with a Theory of Meaning*, Minneapolis: University of Minnesota Press.
Uhlmann, Anthony (2011), *Thinking in Literature: Joyce, Woolf, Nabokov*, New York: Bloomsbury.
Van Suchtelen, Guido (1987), 'Nil Volentibus Arduum: Les amis de Spinoza au travail', *Studia Spinozana* 3: 391–404.
Vasari, Giorgio (2008 [1550]), *The Lives of the Artists*, Oxford: Oxford University Press.
Viljanen, Valtteri (2011), *Spinoza's Geometry of Power*, Cambridge: Cambridge University Press.
Viljanen, Valtteri (2014), 'Spinoza on Virtue and Eternity', in M. J. Kisner and A. Youpa (eds), *Essays on Spinoza's Ethical Theory*, Oxford: Oxford University Press, pp. 258–71.
Viljanen, Valtteri (2015), 'Spinoza's Essentialism in the Short Treatise', in Yitzhak Melamed (ed.), *The Young Spinoza: A Metaphysician in the Making*, Oxford: Oxford University Press, pp. 183–95.
Wainwright, Oliver (2015), 'Revealed: How Developers Exploit Flawed Planning System to Minimize Affordable Housing', *The Guardian*, 25 June 2015, http://www.theguardian.com/cities/2015/jun/25/london-developers-viability-planning-affordable-social-housing-regeneration-oliver-wainwright (accessed 30 June 2015).
Watt, Ian (2001), *The Rise of the Novel: Studies in Defoe, Richardson and Fielding*, Berkeley: University of California Press.
Whiston, William (1702), *A Short View of the Chronology of the Old Testament: And of the Harmony of the Four Evangelists*, Printed at the University Press, for B. Tooke.
White, Stefan (2014), 'Gilles Deleuze and the Project of Architecture: An Expressionist Design-Research Methodology', PhD thesis, University College London.
Wilson, Margaret (1999), '"For they do not agree in nature with us": Spinoza on the Lower Animals', in R. J. Gennaro and C. Huenemann (eds), *New Essays on the Rationalists*, Princeton: Princeton University Press, pp. 336–52.
Wolfson, Harry Austryn (1961 [1934]), *The Philosophy of Spinoza*, Cleveland and New York: The World Publishing Company.
Yong, Ed (2016), *I Contain Multitudes: The Microbes Within Us and a Grander View of Life*, New York: Ecco.

Zimmermann, Rainer E. (2010), *New Ethics Proved in Geometrical Order: Spinozist Reflexions on Evolutionary Systems*, Litchfield Park, AZ: Emergent Publications.

Zirker, Herbert (1997), 'Horse Sense and Sensibility: Some Issues Concerning Utopian Understanding in Gulliver's Travels', *Swift Studies* 12: 85–98.

Index

adequate ideas, 7, 9, 40, 51, 62–3, 125–6, 140, 142, 148–9, 151–2
affects, 6, 28, 39, 41–3, 67, 79, 88, 98–9, 108–9, 131, 135, 139, 144–6, 149, 153; *see also* passions
agency, 2, 42, 94, 97, 109, 111, 113, 117–19, 139
almshouses, 109, 113–14, 117, 124
animals, 2, 35, 57, 59, 69, 93–4, 97, 101, 105–6, 144, 166
architecture, 1, 3–4, 6, 9, 89, 91, 102–4, 106–7, 109, 111–14, 121, 123, 125–6, 133–41, 147–8; *see also* buildings; housing
Aristotle, 37, 41–3

beatitude, 2, 141–3, 146, 151, 153
bodies
   architectural, 96, 137–9
   composite, 35, 61, 66, 77–8, 86
   finite, 10, 14, 20–1, 23–6, 31–3, 35, 37, 61, 160
   human, 2, 10, 27–8, 30, 40, 50, 66–8, 78, 96–7, 110, 112, 147
   parts of, 21, 26, 62, 66–7, 77–9, 125, 130–1, 152
Brünner, Margit, 142–4, 146–9, 153–4
buildings, 3, 9, 90, 94–8, 101–3, 105–6, 125, 133–8; *see also* architecture; housing

citizens, 68–9, 71, 104, 139
common notions, 63, 100, 145, 148, 150–2
communities, 2, 48–9, 68–71, 73, 117–18, 164
conatus, 33, 42–3, 74–6, 78–80, 84–7, 97–8, 109, 111–12, 116, 143
consciousness, 97, 141, 159–61
Conway, Anne, 54, 59–60

Deleuze, Gilles, 2–3, 74–9, 81–8, 100, 105, 125–35, 138, 141–53
Descartes, René, 19, 63, 76, 93, 126, 130, 133, 157, 165
design, 3, 91, 102, 109, 111–12, 114, 116–17, 121, 135, 137

Edwards, Jonathan, 54–6
emotions *see* affects
environments, 3, 29, 74, 89, 91, 98, 104, 118, 121, 142–3, 145–9, 152–4
equality, 54, 61–3, 65, 69, 71–3, 84, 108, 118, 123
essences
   eternal, 4, 63–4, 151, 161
   formal, 7, 11–12, 15, 17
   modal, 11, 74–7, 79, 81, 84–6
eternity, 5, 34, 64, 153, 161–2
Euclid, 8, 11, 15, 61, 63–4, 66–7, 72, 108, 159
Evans, Robin, 125–6, 133–8

fiction, 134, 150, 157–8, 167
freedom, 1–3, 100, 104, 168

geometry, 2–7, 11, 14–15, 63–4, 72, 108–9, 111

harmony, 2, 37–9, 46–60, 62, 69, 71, 73, 91, 97, 112, 118
Hobbes, Thomas, 69, 72–3, 103, 162
housing, 104, 109–10, 112–14, 116, 118–21, 123
   affordable, 109, 118–22
   social, 117, 119, 121, 123–4
housing welfare, 108–9, 111, 113, 115, 117, 119, 121, 123
human nature, 41, 50, 108, 165, 167

Ibn Tufayl, 155, 158–62, 168
imagination, 1, 46, 48–51, 54, 57–8, 80–1, 87, 109, 111–12, 132–7, 150, 160, 167
immanence, 90–2, 128, 132–3, 137, 142–3, 145–7, 153–4
individuation, 44, 81, 95, 98, 108–9, 111, 129
infinity, 33–5, 40, 42–4, 61, 95, 98, 103

joy, 67, 80–1, 98–9, 131–2, 142–3, 148–9, 151–3

knowledge
  first kind of *see* imagination
  second kind of, 61, 145, 150–1, 159–61
  third kind of, 132, 145, 151–2, 159–61

Leibniz, Gottfried Wilhelm, 22–5, 53–4, 58, 76, 165
love, 35–7, 43, 54, 59, 99, 156, 167

Macherey, Pierre, 2, 72, 74–5, 79–84, 86–7

modes
  finite, 14–16, 32, 61, 75–9, 81–8, 131
  infinite, 10–11, 14–17, 145
motion and rest, 10–11, 13–14, 16, 26–7, 30, 33, 35–6, 38–9, 61–2, 65–8, 71–3, 95, 144, 150, 154

parallelism, 19–20
passions, 26, 37, 39–40, 50, 75, 78–82, 84, 86–7, 100, 131–2, 149–52; *see also* affects
perceptions, 39, 133, 135–7, 162
politics, 2, 103; *see also* communities
power to act, 2, 28–31, 74–6, 78–81, 84–8
proportion, 1–4, 33–5, 37–41, 43–5, 47, 54–6, 61–2, 66–7, 73–8, 80–2, 141–3, 149, 153–5, 159–60, 162–5

Ramond, Charles, 74–6, 79, 82–3, 86, 88

sadness, 35, 80–1, 98–9, 131, 143, 152–3
Swift, Jonathan, 2, 155, 157–8, 162–8

wellbeing, 1–3, 33, 35, 43, 103, 109, 111–14, 116–21, 123

EU representative:
Easy Access System Europe
Mustamäe tee 50, 10621 Tallinn, Estonia
Gpsr.requests@easproject.com